Sustainability Inverted

American and Comparative Environmental Policy

David M. Konisky, Megan Mullin, Kate O'Neill, series editors

A list of titles in this series appears at the back of the book.

Sustainability Inverted

How Environmental Policies Control People

Jin Sato

The MIT Press
Cambridge, Massachusetts | London, England

The MIT Press
Massachusetts Institute of Technology
77 Massachusetts Avenue
Cambridge, MA 02139
mitpress.mit.edu

Originally published as *Hanten suru Kankyoukokka: "Jizokukanousei" no Wana wo Koete* by The University of Nagoya Press, Japan

The MIT Press would like to thank the anonymous peer reviewers who provided comments on drafts of this book. The generous work of academic experts is essential for establishing the authority and quality of our publications. We acknowledge with gratitude the contributions of these otherwise uncredited readers.

This book was set in Stone Serif and Stone Sans by Westchester Publishing Services. Printed and bound in the United States of America.

Library of Congress Cataloging-in-Publication Data is available.

ISBN: 978-0-262-05343-3

EU Authorised Representative: Easy Access System Europe, Mustamäe tee 50, 10621 Tallinn, Estonia | Email: gpsr.requests@easproject.com

For my parents

Contents

Preface

Why do environmental policies often produce effects that are the opposite of what they claim to accomplish? I attempt to answer this question through case studies from various parts of Asia and provide potential solutions. Most Asian countries have significantly increased their emphasis on sustainability in recent years—regarding budgets, personnel, and legal frameworks—with substantial support from international organizations, and these efforts have been underway for at least thirty years. Despite all this commitment, reports on the state of the environment continue to indicate that "some longstanding environmental problems are becoming increasingly serious" (ASEAN Secretariat 2023, xxix).

I employ the concept of "inversion" to develop a systematic understanding of this failure. Much of the failure arises when environmental and natural resource policies affect local communities in ways that alienate them from the goals of these policies, whether the policies aim for conservation or resource utilization. "Inversion" occurs when essential local collaborators become adversaries in response to a well-intended environmental policy, ultimately undermining the policy's objectives. Examining the causes of this inversion reveals why so many well-meaning policies fail to achieve their goals, even as they continue to receive international support.

Inversion is similar to but different from what Rob Nixon has termed "slow violence," meaning "delayed destruction that is dispersed across time and space" caused by climate change and other gradually unfolding environmental catastrophes, because they are "typically not viewed as violence at all" (Nixon 2011, 2). Inversion is harder to detect than slow violence since it is often caused by the "solutions" and not by the problems. Solutions are usually interrogated in terms of their effectiveness in solving the problem, not as possible causes of problems. This is why we need to investigate how environmental policies identify and deliver solutions to the "problems." While our attention is usually focused on environmental sustainability, the very policies designed to promote it may inadvertently create an even more serious problem that undermines the social foundation necessary for success.

I argue that foreign donors and interested helpers from outside should move away from a naive faith in "environmental policy." Instead of focusing solely on the state, we should emphasize the decentralized experiences of local intermediaries. We need policies that empower these intermediaries to cultivate in situ long-term relationships with the local environment, thereby promoting stewardship and, consequently, sustainable practices. The examples discussed throughout this book—ranging from pollution in Japan, forestry in Thailand, fisheries in Cambodia, and irrigation in Indonesia—reveal a clear pattern of inversion that offers essential insights into alternative solutions. Improvements in short-term environmental indicators, even if they are significant, should not blind us to the realities faced by local communities and individuals who are continually striving to improve their daily lives.

Technical research examining the decline of natural resources and the increased risk of climate change has grown significantly and is fundamental to addressing global and existential questions regarding the future of humanity. Such studies have provided evidence and impetus for centralizing planning toward optimal efficiency. However, such environmental policymaking has immediate effects on local populations, impacting their livelihoods and their relationships with the state. The lived experiences of local people are important in their own right and possess both manifest and latent potential to shape the future course of policy. The livelihoods of populations in the least developed countries are among the most dependent on the natural resources surrounding them (Ongo Nkoa et al. 2024).

There are two types of environmental policymaking in mainstream thinking. One approach relies on economic incentives, which involve nudging society toward more environmentally friendly practices by using market mechanisms. This includes encouraging the development of new technologies through subsidies and reducing pollution through taxation. This method has proven fairly successful in developed countries, as some sectors have improved their environmental performance. However, some companies have relocated polluting factories to developing countries while retaining ownership, making the impact of their production activities less visible.

The second approach involves a powerful centralized state enforcing regulations. Developing countries often adopt this type of environmental policy. According to the Policy Instruments for the Environment database by the Organization for Economic Co-operation and Development (OECD), non-OECD countries tend to implement more regulatory policy instruments, such as taxes and fees (OECD 2023a). For instance, to control water pollution, a country might issue strict regulations on the use of certain chemicals and utilize police authority to monitor the release of harmful substances into the environment and punish offenders. However, this top-down approach does not inspire individuals to change their behaviors, and any subsequent changes may not be sustainable. Such policies are unlikely to succeed when public trust in the government is low and corruption is prevalent.

I propose a third approach that allows local people to feel a sense of ownership over their immediate habitat. Treating local communities and Non-Governmental Organizations (NGOs) as separate from the market and the state is unrealistic. The ideal of empowering communities has fallen short of expectations in light of the growing dominance of both the market and the state. Instead, I suggest focusing on the nature of *interdependency* between community groups and outside forces, as well as the ways in which members depend on each other within their communities (see chapter 6). In this book, I aim to rediscover the value of interdependency and intermediary groups as agents for changing dependency relationships.

Any policy aimed at improving the natural environment—whether through cleaning up air and water or reducing carbon dioxide emissions—will inevitably affect the societies that inhabit that environment. However, attention has generally been focused on how these policies lead to physical changes in the natural world as opposed to the sociopolitical impacts. Climate shocks increasingly damage the ecological conditions of land and forests through fires, storms, floods, and droughts, forcing people to migrate from their locations. While it is essential to support these individuals, we must also assist local residents who manage to adapt within their communities. They often serve as the most reliable "sensors," utilizing their local knowledge and providing first aid in response to environmental damage. However, as states have strengthened their environmental policies to protect the physical environment, they have also overextended the reach of these policies, thus undermining the ability of local residents to be "sensors." Environmental policies in Asian countries are no longer limited to managing the environment; they have expanded to managing entire societies and people *through* the environment.

When countless aspects of local life are micromanaged by various layers of state apparatuses in their pursuit of the moral imperative to protect the environment, people's relationship with nature can quickly sour. Despite institutional and technical preparedness, the natural environment in most Asian countries is deteriorating at alarming rates, a trend common to all developing nations. Previous studies have not given sufficient thought to the potential impact of this transformation through environmental policy.

People's lives are inextricably intertwined with the power dynamics and politics of environmental regulation, which aims to improve the management of the natural resources they rely on. When evaluating environmental policy, it is essential to ask not only, "Has the water and air become cleaner?" or "Has the forest expanded?" but also, "What effect does this environmental policy (and the science behind it) have on people?" This book begins by posing such questions, revealing why these considerations are crucial.

Hints of "inversion" germinated in my mind during my fieldwork in the 1990s. I was motivated to explore the idea presented in this book after I witnessed how groups of ethnic minorities in rural Thailand were forcibly relocated from their rich forests in the name of protecting biodiversity (Sato 2000). I sought out additional examples to identify the mechanisms at work and consider the extent to which my earlier observations in forest conservation in Thailand could be generalized to other resources and places. This led to studies in state-engineered water irrigation in Indonesia, the distribution of fishery resources in Cambodia, and the fate of communal lands in Thailand after the great tsunami caused by the Indian Ocean earthquake of December 2004.

The more I examined these examples from different countries, the more apparent the patterns became. They redirected my attention from the objective or intent of the policy to its processes. At times, the very *process* of implementing a policy aimed at controlling resources disrupts local people's sense of ownership, thereby disincentivizing their essential participation in conservation efforts.

In this context, inversion refers to the phenomenon where potential local collaborators become adversaries, creating relationships that undermine the very objectives the policy seeks to achieve.

Despite the risk of this ironic turn, the expansion of state power is often accepted and even welcomed by policy elites, while the phenomenon of inversion tends to be overlooked. This oversight is largely attributed to urban residents who, not experiencing the pitfalls of environmental policy, righteously endorse environmental protection and support local communities. However, scrutiny of the processes and failures of such interventions is frequently weak or absent. Although social impact assessments are often included in the environmental impact assessments required for large-scale development projects, they are primarily conducted in advanced countries that prioritize the culture and rights of local people to minimize community damage (Alomoto et al. 2022). In contrast, when local people in developing countries collectively "agree" to these projects, it is difficult to fully understand the sociopolitical ramification on the communities involved.

This book compares the experiences of multiple countries in Asia that are currently undergoing top-down environmental policies and examines the perspectives of local people regarding the effects of these policies. It aims to develop new insights into why environmental policy often fails and discusses how such inversion can be mitigated.

Numerous accounts of conservation projects that come at the cost of local sacrifices have been documented on a case-by-case basis, particularly in Southeast Asia (Pye 2023). Within the forest sector, a key issue has been the conflict between indigenous peoples and protected areas (Milne and Mahanty 2015). The concept of "green extractivism" is now being discussed globally, encompassing the African and Latin American continents (Bruna 2023; Dorn et al. 2022). Evidence indicates that some climate adaptation projects inadvertently reinforce, redistribute, or even create new sources of vulnerability due to existing inequitable sociopolitical relations

(Eriksen et al. 2021). Policies aimed at sustainability transition may generate new injustices and vulnerabilities, prompting many scholars to advocate for a "just transition" (Wang and Lo 2021).

This book presents a different narrative from the previous efforts that problematize the question of power, explicitly focusing on the political impact of environmental policies in Asia, where economic development takes precedence. This developmentalism is underpinned by an emphasis on self-reliance and an independent attitude, which downplays the significance of interdependence—a critical factor in achieving sustainability (Maemura 2024; Ferguson 2015). Backed by a private property system that guarantees returns on investment, participants in capitalism have largely forgotten the importance of our interdependent world, especially at the local level. The book concludes by reevaluating the virtue of interdependency through an analysis of intermediary organizations that have been devalued and are often even viewed as a threat to modernization and the development discourse that gave priority to individuals detached from communities. The key to addressing the inversion lies in rebalancing the disrupted relationships of interdependence and empowering individuals through various intermediaries closer to the ground.

The increasing impacts of natural disasters and climate change around the globe have created a sense of planetary urgency, justifying the accumulation of more power by the centralized states and more regulation by market-based approaches. In such circumstances, it is easy to become convinced by policy proposals suggested by global experts and lose sight of local alternatives. The state's power to control the natural environment differs from the conventional sources of power based on the military and economy. It is more latent and invisible. We, therefore, need a different strategy to explore alternatives.

I wrote this book for those who have wondered why we have made so little progress in environmental policies and for those seeking solutions beyond our conventional reliance on the market and the state. This book is for anyone—academics, think tank policymakers, intermediary groups, government staff—who share the broad goals of environmental conservation and sustainability and are looking for solutions that can help to achieve this in concert with the sustainability of local livelihoods. Ultimately, the book is intended for people committed to serving the voiceless majority living in the countryside of developing countries who continue to depend on nature and community amid the increasing pressure from the market and the state to become more self-reliant and "independent."

Acknowledgments

Writing a book is a process of forming a relationship with others. My experience publishing several books in Japanese has made this clear. This book, which is my first single-authored book in English, owes special gratitude to many, because it extends far beyond the usual academic community with which I had the pleasure of being acquainted. I must acknowledge all the individuals I met in Japan and abroad as well as those who made such international encounters possible.

I would not have imagined pursuing graduate studies in the United States without my high school exchange experience as an AFS (American Field Service) student in Buffalo, New York, from 1985 to 1986. I wish to thank my host family, especially my host parents, Bob and Norma Henderson, for opening up an opportunity to the wider world and for allowing me to be part of their family for the past forty years as their Japanese son. Hosting a foreign high school student for one year would not have been easy, and I appreciate the love and kindness they have continued to offer me. "Jin, when is your English book coming out?" This is the question they have been asking me for the past several years. I can now proudly respond, "Here it is!"

A more direct inspiration for this book came from my fieldwork experience in Thailand as a graduate student during the late 1990s. This was made possible by Mr. Kato Hiroshi and Mr. Sato Masaki, who were in Bangkok at the time and guided my first village experience in the northeast of Thailand. Unfortunately, neither of them is with us anymore. I mention their names here to remember them and express my gratitude for introducing me to Thailand which became my central research site over the thirty years.

During my extensive fieldwork in my PhD years in the late 1990s, I was fortunate to be hosted by Mr. Bunchuai and his family in the remote mountains of Uthai Thani Province. Dr. Somsak Sukwong, the father of community forestry in Thailand, acted as my advisor. Drs. Vitoon Viriyasakultorn and Weerachai Nakviboonwong helped me with all aspects of research implementation, guiding my work on forestry politics. During this fieldwork, I learned that forest conservation, which I had previously thought was a benign policy welcomed by all, had severely constrained the livelihoods of forest dwellers. This book originated from that village experience.

Living in a remote Thai village and other places outside my comfort zone encouraged me to be aware of myself in the position of a "foreign" minority. This occasionally gave me a sense of alienation but also provided opportunities for self-reflection and appreciation of different ways of living and thinking. Through a series of such encounters abroad, I developed the habit of "looking at things from the other side," which has become an undercurrent in this book.

I feel fortunate to have received a liberal arts education at the University of Tokyo, which emphasized "late specialization" at a time when most universities in Japan required students to choose their specialization starting in their freshman year. It was at the Komaba campus of Tokyo University that I decided to devote my career to foreign aid and development. I still remember my visit to Professor Ken'ichirō Hirano, who was my academic advisor, when I asked him for a reference letter to apply to Harvard Kennedy School for my master's degree in public policy. As a Harvard-trained historian, Professor Hirano asked me in a skeptical tone, "Can you read a lot?" He was right. The amount of reading required was far beyond my capacity. However, soon it made me realize that quantity is necessary to achieve quality. My coadvisor, Professor Ishi Hiroyuki, taught me the importance of visiting and talking to people who live in places where the question of interest is most prominently expressed. Revisiting the Karen village in Western Thailand with Professor Ishi and his students, ten years after my fieldwork, remains one of my fondest memories.

My academic encounters in universities abroad further helped me locate the meaning of what I saw in Thailand and other parts of the world. Taking courses with and reading the works of Professor Amartya Sen as a graduate student at Harvard and spending a postdoc at the Agrarian Studies Program of Yale University with Professors James Scott and Michael Dove gave me the foundation of how to define a research problem. These great minds represent a range of academic disciplines—economics, political science, and anthropology, respectively. Exposure to a multidisciplinary environment, and I should say, the very best from each academic tradition, allowed me the wider scope to understand the problems of development, a field that does not belong to any particular discipline.

Much of the original version of this book was published in Japanese as *Hanten suru kankyō kokka*, published by Nagoya University Press in 2019. An English translation was arranged by Academic Language Experts, led by Dr. Nele Noppe (University of Leuven). Nele's nuanced translation provided the basic platform for this work, and I am grateful to Nagoya University Press for allowing me to reproduce the material. This book, however, is not a technical translation of the Japanese version. Following a careful review and suggestions from the series editor, I decided to drop some original chapters and add new ones, updating all chapters with additional fieldwork and literature surveys. I should also note that earlier versions of chapter 3 and chapter 4 were published in *Comparative Studies in Society and History* and the *Journal of Development Studies*. I thank my coauthor, Dr. Thol Dina, for collaborating with me on the Cambodian case studies. I also wish to thank the organizations and individuals who permitted me to reproduce the figures and photographs for this book.

In reshaping the English manuscript, I benefited from feedback provided during my talks at Rutgers University, Princeton University, and Columbia University in the United States from the year 2020 to 2024. I also appreciate the comments I received from my presentations at Chiang Mai University, Thammasat University, and Walailak University in Thailand. In Japan, I had the opportunity to discuss my work at the Research Institute for Humanities and Nature in Japan in 2019. Special thanks to my colleague Dr. Pattajit Tangsinmunkong for guiding me to key forums in Thailand and reconnecting me with Southeast Asia.

The Institute of Advanced Studies on Asia at the University of Tokyo, my home institution, is a unique and privileged workplace that provided me ample time to conduct research and write. I completed much of the book while I was a visiting professor at the Columbia Climate School in New York. Funding from the Toyota Foundation and the UTEC-UTokyo FSI Research Grant Program was indispensable in completing the research and writing the book.

The following people have helped me gather data, checked the bibliography, commented on the general ideas for improvements, or given me moral support as I pushed through the effort to think and write: Natsuko Akiyama, Minami Enoki, Maya Igawa, Megumi Yao, Keiko Ishizaki, Mai Watanabe, Jingyuan Wu, Nao Konno, Thol Dina, Doeurn Seyha, and Anna Ginsburg. A study group of my former students, *Shutsuryoku Kentō-kai* (Discussion Group on Academic Output), has always been a driver and encouragement. I learned much more from them than I taught them as students.

Some people deserve special acknowledgment. Professor Bill Ascher of Claremont McKenna College has consistently encouraged me and supported my career in conducting research in the United States for the past twenty years. The Fulbright Fellowship at Princeton University in 2010 was particularly pivotal, and I am grateful for his support in getting this fellowship. I did not understand the value of a developmental editor until I met Patty A. Gray, who devoted much of her time and skills to improving the book. Patty gave clarity and consistency to the book, a level I could not have achieved by myself. Initially published in Japanese in 2019, I can confidently say that this is a new book with much stronger articulation than the original version.

Perhaps life is more unpredictable than the climate. I never imagined becoming an academic, writing a book in English, or teaching at elite American universities. These opportunities have come to me thanks to the unwavering support of my parents, Kunio and Michiko Sato, who raised me in Nigeria, France, and Thailand during my childhood, and later sent me to the United States for a year of high school and graduate school. Last but not least, I would like to thank my family—Haruko, Sora, and Azusa—now based in Princeton, New Jersey, for continuing to enjoy the unpredictable in life with me.

Introduction: The Invisible State Power

> In former times, we Japanese wore clothes that were like clerical robes, but nowadays we usually wear uniform-like Western clothes. And that is only to be expected, since we have now become one of the most active and efficient nations in the world and our old peaceful harmonious life is a thing of the past. The basic difference lies in the attitude of a people; whether like the Burmese, to accept the world as it is, or try to change it according to one's own design . . . to try to control nature by their own efforts, or yield to it and merge into a broader and deeper order of things. But which of these attitudes, or ways of life, is better for the world and for humanity? Which should we choose?
>
> —Michio Takeyama, 1966. *Harp of Burma*. Rutland, VT: Charles E. Tuttle Co.

0.1 The Missionaries of the Twenty-First Century

Houaykhing, a relatively isolated village in northern Laos, is the site of a forest conservation project organized by the Japan International Cooperation Agency (JICA) in the early 2010s. This project was implemented to mitigate climate change by protecting local forests, which enhance the world's capacity for carbon dioxide absorption and thereby reduce the rate of global warming. JICA staff worked diligently to implement a range of measures based on the latest scientific developments. They utilized a geographic information system (GIS) to create a detailed map of land use in the area, educated local residents about CO_2 emissions, and shared information on the global importance of afforestation and forest conservation.

JICA staff had high expectations regarding their efforts to convey the global significance of forest conservation. To my surprise, the villagers were barely able to read or write due to the absence of elementary schools in the area. This lack of basic education does not imply that the community is unaware of the importance of fundamental science or global issues. However, I wondered if it would not be more beneficial for foreign donors to prioritize sectors where local people already have strong incentives to engage, such as health care or primary education. Witnessing this, I began to question whether my perception was outdated, reflecting biases rooted in a naive belief in the stages of modernization—progressing from fulfilling immediate basic needs to education and eventually to broader concerns like environmental conservation. As I contemplated this question, a member of the local JICA staff attempted, in a casual tone, to explain the "global carbon cycle" to the villagers.

Developed countries have largely recognized the importance of the concept of sustainability. This has mostly come through bitter experiences as a result of the environmental degradation and pollution that have perpetrated and contaminated their lands and seas, as we shall see in chapter 2. However, the sustainability challenge in Laos has taken a unique form. A project was initiated by international actors on unsuspecting Laotian villagers, who had, for the past century, done an excellent job of conserving their rich natural environment without any external motivation or incentives. Now, they found themselves being lectured by experts from the same developed countries that had thoroughly devastated their own green spaces. This paradox raises a critical question: Who truly needs "aid," and who is supposed to benefit from efforts to conserve natural resources?

Perhaps we should take a step back and examine the forces that brought rural villagers in the Lao mountains together with staff from a Japanese aid agency. The scene is reminiscent of the nineteenth-century missionaries who delivered Christianity to the animist mountain people of Southeast Asia. These missionaries did not travel all the way from Europe merely to proselytize; they also aimed to improve the lives of local people by teaching them how to read and write, providing medical care, and attending to other basic needs. While times have changed, the environmental conservationists visiting mountain villages today bear an uncanny resemblance to those missionaries. These advocates sincerely believe in teaching people in the rural backlands of the world's least-developed countries how to behave as responsible and earth-caring global citizens.

Protecting the environment is not the same as spreading religion: Missionaries and aid workers have very different motivations for traveling to these far-flung rural regions. One of the most interesting shifts has been the changing relationship between developed and developing countries. Similarities in the power imbalance are much easier to see today. Visitors from rich countries come via airplanes and four-wheel drives, with smartphones that enable them to access a bewildering array of knowledge, while most rural hosts do not have such resources. The visitors have a clear agenda for change, while the host villagers struggle to live their daily lives.

On the other hand, what is obscured here is a structure of interdependency that was not present in the missionary dynamics. For a proposed aid project to work, the developed countries are forced to seek *support* from the rural populations in emerging nations—regardless of whether the goal is to plant trees or protect existing forests. Through such activities, advanced countries aim for global sustainability, which they are mandated to achieve. Hidden behind a mask of authority, advanced countries are creating a new interdependency with developing nations, some of which still lack the most basic social infrastructure compared to the contemporary standard. The new "missionaries" do not come to save the souls of the savages; they are here to save themselves. Put differently, countries in different stages of development are forging new relationships because they need each other for different reasons. In previous centuries, colonizing powers needed the resources provided by the nations under their control. The present-day relationship makes it unclear what outsiders seek and who does the taking because the outsider comes as a benevolent "aid donor," not as a colonizer. These preconceptions blind us to the fact that the givers must also depend on the receivers.

From the perspective of the local residents of Southeast Asia, forested mountain regions have long been a sanctuary for communities from the hostile dominant groups of the plains (Scott 2009). Ruling elites in the central plain ignored the faraway mountain enclaves, only turning to these peripheral areas in the late nineteenth century when Western capital began to flow in to support the exploitation of forests and mineral resources. Before they were colonized (except Thailand), few Southeast Asian countries had a fully centralized system of government. The management of forests and mineral resources was left mostly to local rulers (see chapter 3). Modernization resulted in a stronger central government supported by new technologies and systems from Western countries. Consequently, the natural environment of the periphery began to be viewed as a cornucopia of resources, and the role of local people was reduced to that of a labor force for extracting and delivering those resources to the central power.

The ascent of international aid organizations after World War II changed the dynamics of large-scale infrastructure projects. This included ways that peripheral people who had once skillfully evaded state control by distancing themselves from the government began cooperating with (and were often co-opted by) that same government. Over time, the power of the state penetrated every corner of its territory, especially along the roads. Outlying regions were increasingly affected by large-scale development projects mandated by the state.

Recently, international aid, while continuing to support developmental projects, has started to emphasize climate-resilient practices.[1] A poster used by JICA to explain its mission (figure 0.1) provides an interesting visualization of the assumption that developed nations are expected to take the initiative in conducting environmental conservation projects in the developing world. These visual aids help to spread information and bolster awareness among local people and are designed to match the particular circumstances of each area.

Let us examine the poster more closely to uncover the basic premises of the donor. This particular poster, crafted to appeal to Laotians, shows two worlds divided by a river. The developed world, including both advanced countries and cities in countries that are still developing, is on the right. This world is depicted as being dominated by factories and buildings whose economic activity not only pollutes the atmosphere but also emits large amounts of carbon dioxide, contributing to global warming.

On the left side of the river, there is a vastly different world: the lush green countryside of Laos. Abundant forests absorb carbon dioxide, and with no factories to produce industrial waste, the sky is clear and blue. This world has no human-made structures. People cradle seedlings in their arms and plant them with great care.

The poster's primary focus is the *encounter* between the two worlds. A group of representatives in formal-looking suits and ties from the right faces a group of Laotian villagers from the left. They appear to be confirming their intent to cooperate for mutual benefit. The gentlemen on the right applaud the new partnership, while the villagers express their gratitude by joining hands. Overall, it

Figure 0.1
The world on the left and the world on the right: *Fair Distribution of Benefits. Source*: Grassroots Capacity Building for REDD+ in Asia project and RECOFTC, Bangkok.

looks like the conclusion of a business agreement. In a sense, this poster accurately conveys how people in developed countries see their involvement in climate-change mitigation projects.

The poster also raises questions. Why are the representatives from these two worlds, which are at utterly different stages of economic development, meeting in the Laotian countryside to begin with? When the Laotian government pushed the country toward economic advancement, rural populations were either left behind or victimized as a result of large-scale infrastructure projects. Why, then, are these villagers suddenly allowed to take center stage in official development assistance (ODA) projects?

We can approach this question by considering what is *not* shown in the poster. In reality, the idyllic Laotian countryside depicted on the left is dotted with major infrastructure projects, such as dams. Some rural regions in Laos have to confront environmentally risky human activities that have long since been outlawed in developed countries. For example, consider the Nam Theun 2 Dam, which became fully operational in 2010. Construction began in 2005 with the support of organizations such as the World Bank. Over six thousand farmers were forcibly relocated to make room for the dam, attracting fierce criticism from NGOs and citizen groups. Laos has few goods suitable for export, and hydroelectric power is therefore seen by the government as a crucial source of foreign currency, with more than half of the electricity generated being exported to neighboring Thailand (Asian Development Bank 2019, 25). Whether such projects truly benefit the people of Laos is questionable. Dams increase shipping traffic on the Mekong River, adversely affecting fisheries resources. The large-scale development of the Sepon mine, which produces copper and other metals, caused ecological damage by polluting the water. The collapse of Saddle Dam on July 23, 2018, resulted in widespread death and destruction (Omar and Kumar 2022).

Another significant reality is overlooked by the poster: who owns the forests earmarked for protection and the land to be newly replanted. In Laos, as in other Asian countries, most afforestation projects have involved mobilizing farmers to plant saplings on surplus government-owned land. It is also common for companies to lease state-owned land for enormous rubber trees and oil palm plantations. This is not a form of afforestation that is likely to bring any significant benefit to local people, except for those who are hired there as workers. Often, in setting up these plantations, the companies drive villagers from their traditional farmlands (Bissonnette and De Koninck 2017; Schoenberger et al. 2017). It is difficult to grasp the true impact of forest conservation and development policies if we focus solely on their effects on the woodlands.

This poster, therefore, does not stand up to scrutiny, nor is it meant for such analysis. However, as we discuss in more detail in chapter 1, these simplified depictions can reflect the "real" perception of the donors; they have a widespread impact precisely because they are so uncomplicated, working to subconsciously reinforce people's outlook on the world. Once such outlooks have taken hold, they form the foundation for government policy, and policy can impact reality. The poster also shows—and to some extent glosses over—a supposedly new relationship between the rich and poor worlds. Colonizers once treated their colonies as suppliers of raw materials. Today, the new awareness that we are all stewards of a shared Earth compels us to develop measures to protect nature globally. In this context, developing countries are no longer merely useful exchange partners; they are indispensable allies in the survival of advanced nations. For locals, however, environmental policies are a mixed bag at best and a heavy burden at worst.

0.2 The Emergence of State-Led "Inversion"

0.2.1 Variations of Inversion

"Time and again throughout history, it was precisely the state's protection of the forest that turned locals into enemies of the woodlands" (Radkau 2008, 307). This is a counterintuitive claim to those who care about nature. Many tend to feel that the state intervenes too little rather than too much concerning environmental problems. Most concerned people would like to see more countries ratifying international conventions on climate change, transferring energy-saving technologies to developing countries, introducing carbon taxes, and subsidizing renewable sources of energy.

Further reflection on Radkau's insight prompts us to problematize the kind of relationships forged between society and the state through state-led environmental policies. Preoccupied with responding to urgent environmental problems, we may invite an even more pressing issue by neglecting the aftermath of state policies on local communities. The foundational issue here is how human society identifies "problems" and how their "solutions" can alter, in turn, the power structure of the societies we live in. Given the repercussions of policies and the resulting need to cover up the negative consequences, perhaps the question is not "how best to solve the problem;" it is rather how to devise the policy in a way that creates the "least problems" or prevents further problems from occurring in its implementation—that is, how a policy can achieve its aims without creating problems that, in turn, require a further policy.

Many governments have already implemented a range of policies to resolve environmental problems. These include designating land as protected areas, transferring electricity generation to renewable energy, using subsidies and taxation to boost the use of eco-friendly chemical fertilizers, and supporting the development of antipollution technologies or nationalizing or privatizing shared resources to protect them. We judge these measures mostly by their effectiveness in conserving and protecting the nature in question. It then becomes difficult to imagine how changes made to nature can lead to changes in our societies. In other words, we forget the simple fact that environmental policies are always implemented *through* human societies, impacting these societies on their way out to the environment.

The process I call "inversion" occurs when environmental protection policies impact local communities in ways that alienate them from the very goals of such protection, and as a result, the local habitat becomes less sustainable. Notably, the effects of inversion do not impact all people to the same degree simultaneously, nor are the agents of inversion always the government. Private companies may advocate for biodiversity conservation, whether as part of their corporate social responsibility (CSR) or to enhance their public image, while intervening in remote areas in search of genetic resources of certain plant species. Their "conservation" activities may restrict local access to areas the company designates as "marketable," thereby limiting resource access for the individuals whose cooperation the companies require (Shiva 2007). The increased market value of certain plants may further invite business interests into the region, resulting in the complete alienation of the locals who might have the best knowledge of how to conserve them. Creating antagonism among the locals is certainly not sustainable in the long term. In other words, an inversion is an unsustainable environmental intervention.

Governments are typically behind the scenes even when the frontline agent is a private company. State efforts to strengthen intellectual property rights in the name of protecting biodiversity often align with the interests of the business sector, which seeks to commodify biodiversity. The enforcement of intellectual property rights can restrict access to genes and other potential bio-resources for other companies, and it can also undermine the rights of communities that once had free access to local resources (Shiva 2015). Attempts to leverage the traditional customs of local populations in marketing nontimber forest products have led to outside companies taking control of the exploitation of those goods (Dove 1993). When the right to exploit resources falls into the hands of external parties, local communities are likely to show little interest in protecting those assets. Intellectual property rights developed by the state appear to prioritize economic interests over the protection of biodiversity. Consequently, the natural environment becomes a means of concentrating power. A fair intervention in this context would require a clear prioritization of rights over economic benefits.

We can observe somewhat related variations of inversion in the case of pollution. As an excellent documentary film, *The Warriors of Qiugang,* demonstrates this from China, there are many examples of chemical factories evading state inspection and secretly discharging pollutants into rivers, a practice that allows such factories to continue operating but also harms the inhabitants of the area (Yang and Lennon 2011). The degradation of trust in superficial environmental safeguards ends up losing the very people who are needed to support the existence of such inherently polluting factories (Zhang et al. 2024).

In addition to private sector involvement, local communities can be mobilized to manage certain natural resources, such as forests and pastures, through incentives offered by the state or international NGOs as part of community-based resource management. While these models of resource management are generally well-intended, they can also privilege access for some while restricting it for others (Hall et al. 2010). For example, private landowners or community members might designate certain areas as "ecotourism sites," excluding nonmembers of the community. Ecotourism serves as another example of a private-sector intervention initiated by external parties that may lead to unintended consequences. Many of these interventions are supported by local governments, which grant licenses to select individuals or groups. However, most external support projects are short-lived, often lasting only a few years. When the project ends and financial resources dwindle, local incentives to cooperate and sustain the project also diminish. This pattern of inversion is frequently triggered by nonresidential NGOs, even when their intentions are positive. Outsiders can inadvertently create dependency on outside resources, which are withdrawn once the project timeline concludes. Consequently, a variety of agents contribute to the emergence of inversions, with the state acting as the primary enabler. For this reason, I focus on "state-led inversion" as the main subject of my investigation.

In all these patterns, the people most likely to be affected are those living in areas targeted for strict conservation projects: communities living near rich forests or rivers that become the target of stringent protection orders, workers in factories that are subjected to antipollution regulations, and people in disaster-prone areas where climate-change adaptation measures are put into practice. Changes in the incentive structure create distance between local people and the resources in question.

Environmental policies, therefore, can range from relatively straightforward management of the natural environment to more nebulous reorganization of human societies *by way of* environmental management, such as imposing new rules and regulations. Those interventions can increase existing inequalities and may cause the very local people whose cooperation is imperative in the success of the environmental policies to become reluctant to cooperate, which then accelerates the damage to the environment.

It is now customary to view the dilemmas of environment and development as a conflict between nature and humanity. Modern societies have encroached further upon the environment, triggering confrontations among different population groups and leading to unintended consequences. If protecting ecosystems were simply a matter of imposing restrictions on pollution-prone companies, terms like "inversion" would be unnecessary. We generally accept that the freedom of some individuals must be restricted for the public good. Furthermore, people in democratic countries can challenge regulations. However, the situation is more complex for rural populations in developing countries whose lives are disrupted in the name of conservation and development of natural resources. They have limited opportunities to contest regulations, and few outsiders recognize their plight, especially since such measures are often labeled as "assistance." It is for this structural reason that we need to illuminate the phenomenon of inversion.

Protecting the environment and advancing economic development are two sides of the same coin. "Developmentalism" is a dogma that calls for a state-led reorganization of a country's entire industrial structure to promote economic growth. It tends to proceed faster and more visibly, with stronger support from political leaders, than setting up environmental regulation that can slow down development. Countries that adopt developmentalism are known as developmental states (Nem Singh 2022; Suehiro 1998), and these have been discussed in the same breath with neoliberalism, which advocates for minimal government intervention and maximal free-market enterprise (Wade 2018). Developmental states can be defined as "organizational complexes in which expert and coherent bureaucratic agencies collaborate with organized private sectors to spur national economic transformation" (Doner et al. 2005, 328).[2]

Environmentalism, on the other hand, is a broad social movement aimed at protecting and improving the environment through political action, policy changes, and shifts in cultural attitudes that mainly started in Western countries but is now spreading globally (Guha 2000). Environmentalism has special implications in developing countries where poverty alleviation is still a priority (Martinez-Alier 2014). To fill the gap, developed countries have pledged substantial budgetary amounts to support environmental and climate initiatives in developing countries. According to the OECD assessment report of 2023, $89.6 billion has been mobilized as climate finance for these nations (OECD 2023b). The availability of funds certainly enables the governments of developing countries to take advantage of this opportunity. The overlay of environmentalism on preexisting developmentalism creates a unique formation of environmental states (see pages 18–19, 35–36) in these countries, which subsequently impacts the lives of their residents.

0.2.2 Focus on the State

Analyses of state-led environmental policy in developing countries tend to place significance on the mere existence of institutions. A focus on environmental institutions, however, will obscure the developmental legacies of older state agencies that carry historical constituencies of support and internally conflicting governance mandates (Rea and Frickel 2023). The conventional approach is particularly weak when applied to emerging countries, where well-designed institutions on paper rarely function in reality. To expand our analytical horizon to include developing countries, I explore the extent to which interventions to protect the environment and ensure the sustainability of resources go beyond the natural environment to dominate the lives of local people in that habitat. State-led environmental policies targeting land, forests, water, or climate have extended their reach to human society as a whole. Most developing countries in Asia have significant rural populations with strong connections to their natural environment, and it would not be an overstatement to suggest that many of these remote regions have been subjected to increasingly onerous environmental engagements.

This aspect of interventions cannot be understood simply by looking at government rhetoric or official data. Because there are no objective standards to assess whether a policy is fully "state-led," it makes sense to use a flexible definition and to identify the phenomena from the perspective of those who are exposed to such policies and must go through the change initiated by the outsiders. Even people in the same country may experience the effects of the environmental policy differently depending on their geographical locations, means of subsistence, ethnicity, and so on. Environmental policy is often enthusiastically supported by urban residents and backed by the international community, who do not experience the natural environment in the way that rural residents in developing countries do. To understand how state-led environmental policy has affected a particular place, we need to consider the local context and the characteristics of its natural environment.

Historically speaking, states have unilaterally expanded their involvement in a range of disparate environmental issues from forests and water to soil, air, and climate; energy; and biodiversity. However, existing research has failed to fully capture the extent of the political and social ramifications of these efforts. There are two explanations for this gap in the literature.

First, environmental studies have become increasingly specialized. The divide between natural and social sciences has been particularly detrimental to our understanding of how nature and humanity interact to influence sustainability. Scholarly consensus places the natural environment within the purview of natural sciences. Environmental experts are skilled at reporting changes in the ecological system itself. Consequently, when these conservationists are called upon to conduct environmental impact assessments of development projects often directed by social scientists, they overlook the need to examine how these projects impact people's lives. Likewise, social scientists give little agency to natural objects as a potential factor in shaping how people relate to each other.[3] Just as we examine the environmental impact of development projects, we need to investigate the developmental impact (i.e., sociopolitical impact) of environmental projects.

A second reason for the lack of scholarly analysis on the impact of conservation on people is the result of modern societies prioritizing intent and various forms of planning resulting from such intent. The primacy of intent suggests that only intended consequences are important, so we overlook the unintended ones. The French philosopher Michel Foucault pointed out that even when people know exactly why they do what they do, they rarely understand the consequences of their actions (Dreyfus and Rabinow 1982, 187). To understand the behavior of the state, we need to employ an analogous approach and recognize that the essence of the state can be found not in its aims or objectives but in its consequences.

State-led environmental policy has significant sway over many aspects of our lives. State intervention in human society by way of the environment can take many forms—for instance, the development and proliferation of renewable energy, the establishment of disaster response infrastructure, environmental taxes, land use systems, emissions regulations, ecology and monitoring of climate change, and other applications of scientific knowledge and imported technology.

In short, they encompass a variety of factors from institutional building to funding activities as well as installing new technology and infrastructure. Perhaps more importantly, they influence our values, as exemplified in the state policies on education and religion.

Projects to safeguard resources and protect nature have been introduced all around the globe in countries at all stages of development, rural and urban alike. These projects have now penetrated the minds of individuals who live there. We are facing unprecedented environmental problems, from chemical pollution to climate change and natural disasters. This onslaught has normalized the involvement of police and military forces, for example, in providing emergency assistance to disaster victims and maintaining order in postdisaster areas.

Longstanding issues such as resource depletion, pollution, and illegal waste dumping require our continuous attention; however, these may not be the primary problems. More fundamental are the shifts in power dynamics resulting from government responses to environmental challenges and their social ramifications. An obsessive preoccupation with the fear of resource scarcity can lead people to adopt more conservative mindsets and prompt countries to consider conflict. Additionally, devastating pollution can trigger mass migration, potentially destabilizing the political equilibrium of surrounding regions.

Problems related to resources or the environment cannot be examined without taking their social and cultural aspects into account. Changes in the natural environment are not inherently "problems" just because they are particularly severe or because they proceed at an alarming pace. Changes in nature become "problems" only when society labels them as such.

0.3 The Chain of Events Leading to Inversion

The impact of inversion may differ depending on how fast or slow it proceeds. We can speak of literal inversion when a policy aimed at protecting the natural habitat results in further degradation. This cause-and-effect relationship usually takes a long time to be noticed. In the 1960s, environmental problems became visible in growth-pursuing states around the world. Rachel Carson's *Silent Spring* (1962) provided a damning condemnation of the use of chemicals in agriculture. It shocked the public into opening its eyes to the consequences of excessive development, setting off grassroots movements to protect nature, especially in the United States.

In Japan, grassroots movements also played a significant role. Citizen-led movements campaigned against pollution early on, even before pollution was conclusively linked to industrial development in the 1950s. They achieved significant successes, including the suspension of plans in 1963 to build a new industrial complex straddling the cities of Mishima and Numazu. Several high-profile pollution trials taught the public how to oppose the government. One such incident was the so-called Itai-itai disease, a painful affliction experienced by the residents of Toyama Prefecture, whose food and water supply was contaminated with cadmium waste from nearby mines. The civil antipollution movement gained support from the Socialist and Communist parties, as well as labor unions, ultimately amassing enough power to compel sluggish local governments to take action. In response to the emerging movements, the Japanese government established the Environmental Agency in 1971.

However, the implementation of environmental policies was not easy. For example, the "Yokohama Method" was a pollution-prevention scheme based on individual contracts between the city of Yokohama and local companies. This regulation was ahead of those stipulated by the national government and was supported by strong environmental concerns from citizens. Between the 1960s and 1970s, this locally devised strategy significantly reduced air pollution and gained widespread praise. However, when the results were scrutinized, it became apparent that certain companies unable to comply with the city's technical standards had simply relocated their pollution-prone activities elsewhere (Kobori 2017). The environmental policies that were considered a "success" in the metropolitan area appeared very different when viewed from the perspective of local residents in the peripheral regions who ended up hosting these pollution-prone activities.

Developed countries like Japan had enough leeway to pursue economic growth first and prioritize environmental protection later. Developing countries today must achieve both at the same time, making their situation different and more challenging. In many formerly colonized countries, "modern" state structures were established during colonization, and the resulting structures of exploitation usually stayed in place even after independence. Rigid political systems and corruption delayed technological innovation, as well as stifling the civil movements and free press that could have raised the alarm about potential environmental problems. Many emerging nations lack the history of social movements that could have generated opportunities for organizational learning, structures to allow the division of labor and coordination in various sectors, and intellectual elites who could provide ideological grounding for citizen action. Unlike the theory of "the stages of economic growth" (Rostow 1960), which suggests that late-coming governments have an inherent incentive to grow alongside their technological capacities, the environmental policies in developing countries are primarily driven and supported by the leadership of advanced nations. Foreign aid and global investments help accelerate the otherwise time-consuming process of having to devise the organizational capacity to manage the transition from focusing on the developmental state to addressing problems of the environment.

What we expect from the government in handling the transition cannot easily be imported from the experience of the advanced nations. The concept of the environmental state is well-discussed in the recent academic literature in the West. Based on case studies in advanced countries, an environmental state is defined as a nation-state that has established a comprehensive set of government ministries and agencies in charge of environmental matters, has enacted basic legislation to bring together disparate laws that affect nature, has ratified relevant international conventions, and is training dedicated experts (Duit et al. 2016).

This definition, however, blinds us to what environmental states do when they emerge in developing countries. We need to examine the nature and behavior of the states themselves rather than trying to import models from advanced nations. Again, we need to take into consideration the fact that the institutions in place rarely respond to the needs on the ground (Sato 2013a). To expand our analytical horizon to include developing countries, I would define the environmental state as one whose interventions to protect the environment and ensure the sustainability of resources go beyond the natural environment to dominate the lives of local people living in that environment. An environmental state in developing countries manifests itself through interventions targeting land, forests, water, or climate that extend beyond the environment per se to cover human society as a whole—and indeed, this is what "environment" means in a literal sense.[4] Most developing countries in Asia have significant rural populations who have a strong connection with their natural environment, and it is no exaggeration to say that all these remote regions have been affected by interventions from an environmental state. When the effects of environmental policies become inverted, leading to the need for new solutions, these "solutions" tend to be drawn from the same urban-based developmental mindset that generated environmental problems in the first place, creating a cycle of self-reinforcing interventions.

Advanced nations had ample time to transition from developmental states to environmental states, albeit in varying degrees. Emergent nations have had to implement wide-ranging environmental policies while their legal and educational systems are still underdeveloped and significant social disparities linger. Peripheral regions enduring the brunt of the impact of these policies are likely to experience inversion.

Developing countries have fewer domestic resources to address the side effects of economic growth, such as environmental degradation and pollution. With support from developed countries, they can formulate, implement, adapt, and strengthen policies specifically targeting environmental issues, thereby expanding the regulatory function of government. The implementation of these policies to protect nature and manage resources contributes to the rise of the environmental state, transforming human society in the process.

Let us compare the experiences of developed countries with those of developing countries. First, development and the environment are mutually interlinked, and the effects are also mutual. Researchers are aware that large-scale economic development projects often lead to demand for environmental policies to mitigate the side effects. What is rarely discussed is that environmental policies can also lead to calls for more development, especially technological measures such as investments in further infrastructure, which can subsequently lead to an inversion of intended outcomes. The displacement of local residents to create large-scale renewable energy farms, such as solar and wind farms, along with subsequent development projects designed to compensate for this displacement, is a typical example.

Second, developed and developing countries have experienced this transition at markedly different speeds. For advanced nations, the process of moving from a policy emphasis on development to an emphasis on the environment took many decades or more.[5] Developing nations today are experiencing an accelerated speed of social change, and the transition period is particularly short in emerging economies of Asia. Unlike those in more advanced nations, which evolved gradually, local communities in the latecomers of Asia have not had enough time to adapt to this rapid transition. The pressure to simultaneously implement both state-led development and national environmental policies creates numerous and often conflicting circumstances that result in unsustainable consequences. This whole process is facilitated through aid and technological transfer from economically advanced countries as well as through the economic logic that justifies offshoring of the polluting industries from advanced nations to developing countries. It is therefore difficult for developing countries to regulate the transition autonomously.

Emerging states go through the sequence much faster for two reasons. First, these countries receive support and backing from international donors, such as the World Bank, to help them manage development programs and safeguard the environment at the same time. Second, these countries lack the forces that would have the power to challenge this acceleration, such as a sufficient variety of intermediary organizations (see chapter 6). As a result, countries still struggling to overcome profound domestic poverty are forced to go through the sequence of development at an accelerated speed. When the burden of environmental protection is added, they risk increasing unresolved disparities among their people.

Table 0.1 demonstrates the evolution of environmental policies in advanced countries (UK, Germany, and Japan) and late-developing countries (Thailand, Indonesia, and Cambodia). I focus on Japan and the latter three countries because they represent a variety of stages, speeds, and manifestations of inversion. Perhaps an additional explanation for why Japan should be included in the case studies is needed. As we shall see, Japan experienced one of the most severe consequences of environmental damage that the world had seen in the twentieth century. As is well documented and also represented in the popular movie *Minamata* (2020), Japan demonstrated the dark side of economic success and how local social movements might be essential in minimizing such damage (as will be discussed further in chapter 2). Japan also became a top donor of foreign aid in the 1980s, replicating the large-scale infrastructure development that helped lift it out of poverty after World War II, particularly in Southeast Asia. This dual position of Japan makes it a unique and important case for examining the long-term evolution of state-society relations concerning the environment.

Table 0.1 shows how these countries' progress in developing the whole package of environmental policies, including the relevant government agencies, compares to the time it took some major advanced countries to introduce similar systems. Both Indonesia and Thailand, for example, began setting up their administrative systems for the environment in the 1970s, triggered by the UN Conference on the Human Environment in Stockholm. By the time of the 1992 United Nations Conference on Environment and Development, also known as the Rio de Janeiro Earth Summit, both countries were developing into states with full-fledged interventions in environmental matters, displaying considerable awareness of the need for conservation as a global issue.

Cambodia, in striking contrast, rapidly cobbled together a system in the late 1990s. Today, the country boasts comprehensive environmental policies—at least at first glance. Cambodia achieved rapid progress because experts from developed countries and international organizations provided on-the-spot support in setting up formal institutions. But just because these institutions exist does not mean that they are effective. In many emergent nations, the actual situation does not reflect the ostensibly solid institutions put in place to protect the environment.

Taking a closer look at the table, one trend is immediately clear. While it took already-advanced countries like Japan about thirty years to travel from a growing awareness of environmental problems to the completion of formal institutions to tackle them, developing countries went through this process in much less time. This difference in speed is emphasized here because the less time a country has to proceed through the sequence, the more difficult it is for societies to respond to the state's environmental policies, and the more likely it is that an inversion of environmental policies could occur. A faster pace often translates into greater suffering for those

Table 0.1
Compressed evolution of environmental states

	UK	Germany	Japan	Thailand	Indonesia	Cambodia
1950s	Water Act (1951) Clean Air Act (1956)	Federal Water Act (1957)	Water Quality Control (1958)			
1960s			Smoke and Soot Regulation Law (1962) Basic Law for Environmental Pollution (1967) Air Pollution Control Act, Noise Regulation Law (1968)	Factories Act (1969)		
1970s	Department of the Environment (1970)	Federal Immission Control Act (1974) German Environment Agency (1974)	Water Pollution Prevention Act (1970) Environment Agency (1971)	Enhancement and Conservation of National Environmental Quality Act (1975) Environmental assessment regulations (1978)	State Minister Office of Development Monitoring and Environment (1978)	

(*continued*)

Table 0.1
(continued)

	UK	Germany	Japan	Thailand	Indonesia	Cambodia
		Federal Nature Conservation Act (1976)	Nature Conservation Law (1972)			
1980s	Town and Country Planning (Environmental Impact Assessment) Regulations (1988)			Ministry of Science, Technology, and the Environment (1982)	Environmental Management Act (1982) Environmental impact assessment system (1986)	
1990s	Environmental Protection Act, environmental assessment regulations (1990)	Act on the Assessment of Environmental Impacts (1990)	Basic Environment Law (1993)	Environment law, mandatory assessments (1992)	Regulation on management of water quality and control (1990)	Law on Environmental Protection and Natural Resources Management (1996)

	UK	Germany	Japan	Thailand	Indonesia	Cambodia
			Environmental Impact Assessment Law (1997)		Start of air pollution program (1991) Environmental Management Act, revised (1997)	Sub-Decree on Water Pollution Control (1999)
2000–			Ministry of the Environment (2001)	Ministry of Natural Resources and Environment (2002)		Sub-Decree on the Control of Air Pollution and Noise Disturbance (2000)

Source: Created by the author.

who have to bear the short-term costs of inversion—for example, having to follow stricter state regulation on resource use, while the long-term costs will be borne by the deteriorating environment. This book focuses mostly on the developing regions in Asia, not because their economic development started later but because the sequence from developmental state to environmental state has progressed extremely rapidly.

The speed has the effect of pushing the important question aside. For example, if a factory that emits toxic substances installs desulfurization units, emissions will certainly fall. Policymakers often favor familiar solutions rather than questioning the very existence of the pollution-generating factory. Environmental policy will always remain subordinate to the development-first agenda that made these policies necessary. A development agenda that necessitates environment policies is likely to trigger inversion.

0.4 Can We Learn from Advanced Countries?

In developed countries, grassroots environmental movements have called attention to and softened the violent impact of developmentalism. In Japan, pollution scandals gradually united disparate groups and institutions—from the mass media and the courts to academics—in supporting local residents affected by environmental problems. This is why the experience of inversion due to environmental policy, which is so prominent in other Asian countries, was not so drastic in Japan. From the 1960s to the early 1970s, Japanese society cultivated organizational structures that enabled people to mount genuine opposition to government policies.

By the 1970s, countries like Japan had laid the groundwork for environmental protection programs. In comparison, emergent nations in Southeast Asia had a late start, simply because these countries were only just entering the stage of a "developmental state." It wasn't until the 1950s that Southeast Asian nations began to draw up national development plans and kick-start economic progress, many with support from the more advanced countries (Tambunlertchai and Gupta 1993). It is easier for politicians to attract voters with "development" that brings material benefits to many: roads and bridges, railways, schools, hospitals, and so on. Environmental measures that could delay such obvious development are much less popular. Thus, it is hard for developed-country solutions to take root in a developing society, especially among the rural populations whose dire need is to improve their immediate living standards.

Another factor regarding the progress of environmental policy in developing nations deserves recognition. When targeting financial aid, developed countries assess a recipient country's progress on various issues, including gender equality, conservation, citizen participation, transparency, and sustainability. However, these are the same issues that advanced nations largely ignored during their own economic development, which shaped their current status (Chang 2002). Ideals such as gender equality and sustainability were not imposed by the governments of advanced countries; rather, they were gradually established through long and arduous negotiations between citizens and their leaders.

As table 0.1 shows, Japan took forty years to establish its environmental protection system. In contrast, emerging countries have had to accomplish this in less than half that time. Nevertheless, from another perspective, developing countries benefit from what is known as the "latecomer advantage," allowing them to leverage advancements made elsewhere. For instance, many countries now require environmental assessments as a standard practice before implementing large-scale infrastructure projects. However, it is important to recognize that if environmental policies are integrated with a strong emphasis on development, a country may continue to operate according to developmental principles. This situation can hinder its ability to address the root causes of conservation problems, even if it has successfully established a system to protect nature.

Much can be learned from examining how environmental policy emerged in advanced countries. However, the most valuable lessons are not about how these countries developed technologies or institutions but rather about how they cultivated the societal capacity to absorb rapid change. Many emerging countries skipped this crucial step in their haste to implement environmental management—specifically, nurturing their societies to accept the necessary policies. While many countries now have systems to assess the environmental impact of development-related projects, virtually none possess the ability to evaluate the social impact of these environmental policies. In this context, halting inversion becomes a lost cause.

Globalization brings together groups of people who have established distinct relationships with their natural environments. These groups are now compelled to find a balance between the differing values and principles that govern their behaviors. Important questions arise, such as why Japan imports timber for construction when nearly 70 percent of the country is covered in forests. Additionally, why can villagers in Southeast Asia be evicted from the lush woodlands where they have lived for generations in the name of conserving biodiversity? Consider the poster at the beginning of this introduction, which depicts a hypothetical meeting between Laotian villagers and representatives from developed countries. This imaginary meeting symbolizes the distortions caused by the varying paces at which different groups have experienced development.

Climate change and extreme weather events are forcing desperate people to leave areas where farms are destroyed or water is scarce. So-called climate refugees are a significant concern for many. Conversely, some individuals maintain a close connection with their surrounding environment. The concept of inversion will have major implications for how we support these local populations as stewards of resources rather than adversaries to their governments. Inversion is becoming increasingly evident across Asia as we observe more state interventions in the conservation and utilization of natural resources.

0.5 Plan of the Book

Chapter 1 outlines the schematic patterns of how natural resource management by the state can turn into a control mechanism over people, with a geographical focus on Southeast Asia. The chapter analyzes how the nature of resources can invite certain types of governance structures and how competition among state agencies further determines the distribution of costs and benefits pertaining to resource governance.

Chapter 2 questions our basic knowledge system that triggers and often sustains inversion. By focusing on the notorious case of pollution in Minamata, the first environmental issue to attract international attention in Japan, I examine how formal science becomes harmful and may invite inversion. I highlight the career trajectory of the nonconformist water engineer, Jun Ui, who exposed and challenged the "knowledge establishment," such as the group of medical scientists at the University of Tokyo. Ui formed his group to press society toward justice and understanding of the local situation. The story of Minamata should be remembered not only as a harsh example of pollution and the experience victims went through but also for the scientists and authorities who were complicit in covering up the cause of the pollution and thus prolonging the damage.

Chapter 3 explores how the seeds of inversion were planted in the early modernization period. Focusing in detail on the historical evolution of resource governance in Japan and Thailand, I illustrate how mining and forestry have served as instruments for state control in the processes of state-building. This chapter addresses the question of why Japan developed an inclusive system of resource governance, while Thailand, another uncolonized country in Asia, pursued an exclusive policy that denies resource access to local people.

Chapters 4 and 5 examine contemporary cases of inversions in Southeast Asia based on my fieldwork in Cambodia and Indonesia. Chapter 4 explores the example of fisheries of Tonle Sap Lake in Cambodia, where we find a rare incidence of the government handing over resource access to local communities. I observe the social and ecological consequences of such an open-access policy. By investigating the underpinnings of these initiatives which are ostensibly meant to empower the community, the chapter uncovers how seemingly benign resource conservation can turn into a tool for political manipulation. Chapter 5 is a case study of irrigation management and how water control techniques mediate the relationship between the state and local people in Southeast Asia. Focusing on the case of Indonesian irrigation, the chapter offers a fresh perspective on how local people establish interdependent relations with the state through water management, often calling for the state to intervene in managing water resources that are indispensable to rural livelihoods.

Chapter 6 discusses potential solutions to inversion. I posit intermediary groups as mitigating agents of state power and inversion. Economic development and modernization were accompanied by the tendency to dismantle a variety of intermediary organizations that were thought to interfere with the centralizing desire of the modern state. Along with the emphasis on individual freedom and wealth, the roles of intermediary organizations as potential protectors of the local environment have been weakened. I argue that the revitalization of these intermediaries is not a nostalgic ideal but rather a necessary scenario to allow local people to thrive in each location without having to feel a sense of isolation or external control in the age of climate and environmental justice.

The concluding chapter synthesizes the above-mentioned cases and explores their broader significance. I argue that the nature of interdependencies, often portrayed as inimical to self-reliant development, can serve as a basis of cooperation—especially among the intermediary organizations that stand between the states and local people. Escaping inversion requires us to appreciate the virtue of interdependency rather than demonizing it.

1
Mechanisms: From Dominance over Nature to Dominance over People

> Thus, through the circumstances of their lives, and the nature of our own wants, all these have been our allies in keeping the balance of nature tilted in our favor. Yet we have turned our artillery against our friends.
>
> —Rachel Carson, *Silent Spring*

How did states come to control the Earth's vast, once-untamed nature? How did this process of control, in turn, affect local communities that had formed relationships with that nature long before the state intervened? Resources such as forests and pastures are inherently local. Nonetheless, state actions impact local populations through overarching policies related to trade, taxes, ownership rights, and infrastructure development. In this chapter, I scrutinize examples from Asia and beyond to explore how the state has exerted its dominance over nature, with collateral effects on local populations. I investigate the typical methods used by governments and the sequence in which policies are implemented. By analyzing patterns of these mechanisms, I explain how the emerging new policy areas are in dire need of a justice-centered analysis.

1.1 Environments as Subjects of Control

1.1.1 The Project of "Legibility"

In many developing countries, the state uses nature to extend its power with little consultation from civil society. Such countries often race through a development process that took well over a century for typical advanced nations to complete, compressing it into mere decades. Caught in this rapid cycle, many emerging nations are ill-prepared to address crises that require the participation of local people. Civil society organizations, even when they exist and function, are also relatively unaware of which aspects of the environment need protection and why.

The late-growing economies of Southeast Asian nations, such as Laos, Cambodia, and Myanmar, exemplify these conditions. Highly reliant on external assistance, these countries turn to foreign companies for help with large-scale development projects. They also seek support from other countries or international organizations for institutional and technological solutions to environmental problems. Such "imported" policies yield unexpected outcomes, highlighting the challenges of applying external solutions in new local contexts. For example, the REDD+ (Reducing Emissions from Deforestation and Forest Degradation in Developing Countries) projects moderated by the United Nations have faced criticism for failing in multiple instances across three continents (Jones and Lewis 2023).[1] This scheme aims to reduce global CO_2 emissions by conserving forests in developing countries in exchange for financial incentives provided by developed countries. Like other international aid arrangements, they tend to neglect the needs of the local people where such projects are implemented. Aside from the uncertain impact on the actual reduction of CO_2 emissions, the effects of REDD+ on peasants and small-scale farmers are more concerning. As organizations such as the World Rainforest Movement continue to report, peasant farming is blamed for deforestation, and traditional forest use practices like shifting cultivation are restricted, while large-scale destruction driven by corporations continues unabated.

The process described above outlines a pathway for the typical transformation of developmental states into environmental states. The concept of the environmental state clarifies one aspect of governmental attempts to expand their authority into environmental arena. This encompasses not only natural resources, but also interactions among people and reshaping relationships and the fabric of society—a process that can lead to inversion. In justifying its slogan, an environmental state may employ various rhetorical techniques to define what it considers to be the "problems." The state will not act on depleted resources or pollution, or implement specific solutions, merely because they are identified as "problems" by a local or by an external group. Rather, the state selectively emphasizes certain aspects of the environment and problematizes them as policy targets, leaving aside the ecological and related sociocultural complexity involved. Understanding the forces at work during this tuning process is essential in analyzing how states invite inversion. In short, my analysis is based not on state intentions for environmental conservation and resource development but rather on the local consequences of the state changing its priority. If a state issues announcements about pursuing a policy of conserving and utilizing natural resources that significantly impacts people's livelihoods, I label it as an "environmental state." I find this concept useful because, unlike the developmental state—which often invites critical perspectives on the repercussions of "development," such as rent-seeking and corruption—environmental policy is often seen as serving a good cause, concerning *things* and not humans, and rarely attracts political scrutiny.

As outlined in the introduction, an environmental state does not simply emerge when a state reaches an identifiable point of no return. The state's systematic involvement with nature begins with the development of sectors like forestry and irrigation. Subsequently, the state expands its activities into other areas, such as sustainability and environmental protection, ultimately evolving into an "environmental state." Many signs indicate the emergence of an environmental state. A country may establish government ministries responsible for implementing comprehensive environmental policies or craft laws to legitimize these new activities. It may also institutionalize regulations and systems of redistribution. Furthermore, a state may cultivate environmental experts to educate the public (Duit et al. 2016). There has been a significant global spread of environmental ministries, particularly in developing countries since the 1990s, where democracy and the inflow of foreign aid facilitate the process (Aklin and Urpelainen 2014). To fully comprehend how an environmental state functions, we must look beyond its administrative systems and consider how the state's environment-focused interventions reverberate throughout society.

When modern nations first sought to identify "legible" targets for government policy, they classified individuals according to race, occupation, economic class, and religion. Similarly, when a state aims to intervene in nature, it must undergo a comparable process to establish legibility (e.g., of certain trees or minerals that provide revenue), ensuring that policy targets can be monitored and manipulated (Scott 1998, 2). Why is this necessary? The "environment" surrounding human societies is perceived through our senses (heat, smell, scenery, etc.), rendering it a poorly defined target for coherent policies. An additional challenge in making the environment a topic of policy debate is that developmental states often have a history of doing the exact opposite. Government officials frequently spend decades asserting that pollution is a necessary evil accompanying economic development, dismissing contamination incidents as mere local issues that lie outside the state's purview. After a state has clung to this narrative for an extended period, it may suddenly change course and argue that environmental policy is an issue requiring intervention at the national level. This is where legibility becomes crucial.

If people increasingly come under state control, is that always a negative development? Not necessarily. For instance, in the context of disaster response, the absence of a centralized command system to distribute goods and services would lead to chaos, inefficiency, and delays. However, under normal conditions, increased state control—especially when it occurs without people's consent—limits individuals' capacity and opportunities to act autonomously in those instances when they believe the state's approach is flawed and requires revision. More importantly, such a top-down approach may prevent state actors from understanding ground-level realities, which are often essential for the successful implementation of any policy.

Governments employ various technologies and systems to reach out to communities to identify problems that justify intervention. Nonverbal representations, such as statistics, maps, aerial photos, and even cartoon images like the graphic shown in the introduction, can facilitate an intuitive understanding of a situation and demonstrate how and where the environment has changed. Governments can also promote citizen organizations to engage with specific issues, their causes, and potential solutions while crafting laws to regulate these groups. Putting a price on CO_2 emissions, a common economic tool to make global warming legible, also has implications for local communities since emission reduction has to take place somewhere on the ground.[2] In the context of developing countries, the distributional impact of carbon pricing is found to have the potential to increase poverty and inequality (Shang 2021). For project-based activities, people residing in the project areas often have no voice in decision-making and no direct claims to the revenue generated from pricing. They are affected by the consequences of project implementation, while policies at the higher level tend to protect the vested interests that determine the direction of environmental policies.

1.1.2 When Politicians Manipulate Nature

While making certain aspects of nature legible targets for intervention, identifying the root causes of environmental problems is much more challenging. This difficulty obscures the burdens that conservation imposes on people, leading to unchecked inversions that leave environment-related policy initiatives open to political manipulation. When these policies are inverted, those who suffer the most are usually those with the least political power.

The global network of commodities influences the behavior of environmental states in different geographical locations. Two processes illustrate this. Between 1970 and 1982, Japan was a major importer of wood, to the extent that some NGOs labeled it an "environmental terrorist." During this period, the Marcos regime dominated the Philippines, and timber exports from that country involved large-scale political patronage, cronyism, and the exploitation of forest resources. The taxes collected from timber in the Philippines, as reported by forestry companies, accounted for only 14.8 percent of the total value of its timber exports (Kuroda and Nectoux 1989). This suggests that a significant portion of logging profits bypassed the Philippine Treasury (Repetto and Gillis 1988). Such overexploitation is unsustainable but continues in the name of economic development that may not be equitable. Ideally, profits from resource exploitation and the natural environment should flow to the government in the form of tax revenue. Then—again, only in an ideal world—these funds would be used to improve the lives of the public in general as well as those who were directly impacted by resource extraction in particular. More commonly, however, instead of enriching the government, this wealth often ends up in the pockets of elites, as seen in Japan's timber imports from Southeast Asia.

Profits may still contribute to the state's power in ways other than taxes. During Suharto's presidency in Indonesia (1968–1998), for example, government officials commonly used forest and oil reserve concessions as political currency. Although logging concessions held significant commercial value, this alone does not explain their importance to Indonesian politicians. These valuable concessions often covered businesses in remote regions, allowing forest exploitation to continue with little oversight. To ensure profitability, the Indonesian government imposed low logging levies on these businesses, and some of the profits were allocated for development projects endorsed by Suharto (Ascher 1999). This type of developmental state arrangement laid the groundwork for the later emergence of environmental states.

Such a setup enabled the Indonesian government to secretly finance projects that would not have withstood public scrutiny in parliament. Kickbacks from logging companies allowed politicians to consolidate their power base without legislative oversight. For instance, Suharto diverted the vast wealth generated from exploiting the nation's forests to the aircraft industry. Additionally, the Indonesian government's decision to set logging levies below market value led to consequences beyond immediate economic concerns, such as accelerated deforestation, as companies were granted concessions far larger than necessary. Massive revenues from oil extraction played a similar role in Indonesia's corrupt system. To secure his power base, Suharto needed to keep the army in check. Oil revenue, like logging profits, bypassed the treasury and flowed into the pockets of the military, enabling Suharto to maintain control over the armed forces.

Developmental states in the past limited their interventions to the development of natural resources and infrastructure for creating wealth and profits. Today, their domains extend to aspects of nature that are essential to the sustainability of human life: air, energy, climate, and water. In the environmental state, regulations ostensibly imposed to limit pollution and prevent overexploitation often become opportunities for pork-barrel politics. Government incursions into new domains—and the reactions these elicit from companies and individuals—can become the seeds of inversion. Perhaps the latest battlefield is in the renewable energy sector, where China and South Korea are leading in technological innovation. However, the renewable energy economy still relies on the extraction of natural resources, including rare earth minerals and metals, which may invite a new form of inversion (Thurbon et al. 2023, 218). The vested interests and the processes that evolved during the developmental state have remained strong in the transition to an environmental state.

1.2 How States Expand Their Domains

1.2.1 When the Management of Resources Becomes the Management of the Environment

Natural resources were once controlled by local people. How did the environment end up under the jurisdiction of the state instead? While we will leave a detailed discussion of this topic until chapter 2, a brief overview of the history of Japan and other Asian countries provides some insight.

When countries in Asia first began their transformation into modern nations in the late nineteenth century, their governments perceived resources like woodlands and mineral deposits as commodities to be exported in order to secure funds for modernizing key equipment for industrialization and empowering the military. In Japan, the Meiji government in the late nineteenth century had no hesitation in auctioning off government-owned forests to overcome budget shortages. Minerals received similar treatment as the government sought to commodify precious metals—from gold and silver to iron and copper—to fuel Japan's modernization. The management of land and fishery resources by the state came later, but Japan advanced to this stage relatively quickly compared to the Western countries; after all, land surveys had been organized since the Sengoku period (mid-fifteenth to late sixteenth centuries) to govern its limited amount of arable terrain.

When the Meiji government sought to strengthen its management of the nation's land in the nineteenth century, it already had access to extensive local information. All land registered with the state was included in a cadastral map, which the government used as the basis for levying land tax. Over time, the state's environmental interventions grew beyond its systems for economic gain, expanding its purview to include broader management of the environment. This increased state involvement is evident in the development of recent climate policies that override earlier regulations on water and air quality, as well as energy supply.[3] In short, as the example of Japan illustrates, state control over the natural environment does not evolve in isolation; as governments broaden their interests, they impose greater controls.

The emergence of an environmental state has flourished globally since the 1972 United Nations Conference on the Human Environment. Beyond the use of individual natural resources, states began to view and manage these resources as a system, taking into account the side effects of development, such as inequality and pollution. The Stockholm conference catalyzed developed and developing countries to establish systems for implementing environmental policies, with Southeast Asian nations following suit. In 1974, Malaysia enacted the Environmental Quality Act, introducing comprehensive environmental regulations. Thailand followed in 1975, while the Philippines established full-fledged legal and administrative structures to manage environment-related risks in the late 1970s, with Indonesia doing the same in the early 1980s. Vietnam, a relative latecomer, did not take similar steps until the 1990s, and Cambodia did not enact an environmental policy until the late 1990s. Both Vietnam and Cambodia faced challenges in enforcing the policies enacted by their newly created institutions, largely relying on support from OECD countries and the World Bank, which provided expert advisors and significant financial assistance.

In terms of governance, the management of natural resources has implications that extend far beyond the conventional understanding of the "environment." It influences the development of infrastructure, the imposition of taxes on specific products, the collection of data on resource distribution, and more. Another example can be seen in table 1.1, which charts the chronological sequence of Thailand's interventions in the environment, illustrating how its environmental state emerged. However, having the institutional setup is one thing, and effective policy implementation is quite another. For instance, Ding and Beh argue that "ASEAN's environmental performance is severely lagging compared to other regions despite evidence of its cohesive and comprehensive efforts to mitigate emissions" (Ding and Beh 2022, 1).

If ASEAN (Association of Southeast Asian Nations) governments are not effectively protecting the environment, what are their environmental policies accomplishing? Since the 1990s, loss of biodiversity, pollution, and deforestation have become increasingly prominent policy issues that cannot be ignored even in developing countries. Government agencies tasked with addressing these crises, however, have largely approached the issues from a conservation perspective. The dominance over people is often an unintended consequence. Countries seeking to control nature are not necessarily planning to increase their control over people. Intentions aside, the results of such interventions often go beyond the government's stated objectives. For example, a conservation project that expands forest acreage may necessitate the displacement of Indigenous people living in the area. When these people relocate, they may resort to cutting down trees in their new location to support their livelihoods. Ultimately, this "successful" afforestation project may exacerbate overall deforestation. The lesson here emphasizes the importance of focusing on the effects of intervention, not the intentions behind them.

Using the example of Thailand, table 1.1 helps us understand a key mechanism of the environmental state: how it converts dominance over nature into dominance over people. The right-hand column in table 1.1 illustrates the social effects of various interventions. When the state attempts to manage a particular resource, it establishes a system based on inclusion and exclusion. Governments often find it convenient to single out a specific group of people as the "cause" of a problem while identifying others as the "solution." Moreover, these stakeholders do not start out as equals. Government interventions tend to adversely affect those who are already disadvantaged in terms of economic power, ethnicity, gender, residence, or language.

Concerns over adverse effects are particularly relevant when discussions shift from tangible, visible resources like soil to abstract concepts that require the state to establish seemingly arbitrary standards. Radioactive contamination and declining biodiversity serve as good examples of more abstract environmental problems. Here, the outcome of an intervention depends on the level of education and awareness among the people involved. Even when an initiative is nominally successful, it may lead to widening disparities between those who set the standards and those who must accept them as given.

Not every country will expand the scope of its environmental interventions in the order shown in table 1.1. Interventions occur at different times and have varying levels of success, depending on local circumstances. In the Philippines, for example, water regulations introduced during Spanish colonial rule were the state's earliest environmental intervention (Magtolis and Indab 2008). Many Southeast Asian countries have a high land-to-population ratio and thus tend to prioritize securing a labor force. They may also postpone establishing private ownership for every plot of land in their territory because the necessary cadaster production would be incredibly complex. In contrast, Japan demonstrates the opposite case: Arable land is scarce, which prompted the central government to get involved in land registration at an early stage (see chapter 2).

Some scholars note that environmental states in developed countries are already showing signs of decline. Environmental sociologist Arthur Mol found that the number of people employed by environment-related ministries and government agencies in many

Table 1.1
Chronological sequence of Thailand's interventions in the environment

Stages in sequence	Characterization of resources	Technologies of intervention	Incentives for intervention
Integration into the global economy (nineteenth century)	Marketable goods (e.g., teak)	Private property system	Revenue collection, dispute resolution
Expansion of domestic production (early twentieth century)	Mobile goods (irrigation water, fish)	Infrastructure development	Production enhancement
Emergence of interest in conservation (mid-twentieth century)	Land, including the goods (forests)	Mapping and statistics, local participation	Revenue security, dispute resolution
Introduction of scientific management (late twentieth century)	Land, including multiple goods (biodiversity)	Species-based map, land zoning, education	Public goods security, international standard
State expansion into cross-sectoral resource management (late twentieth century)	Environmental space with different intensity (air, water, waste)	Monitoring of output and law enforcement	Public complainants, health risk
Further expansion of state interests (early twenty-first century)	Environmental space with similar intensity (climate)	Monitoring of source, education	International standard

Source: Created by the author.

developed nations is stagnant or falling (Mol 2016). However, in less developed countries that rely on large flows of international aid, the influence of the environmental state is still expanding. As they adjust to changes brought about by environment-oriented policies, people often find themselves overwhelmed by the speed at which policymakers target ever-expanding aspects of nature. Take Laos and Cambodia, two of the least developed countries in Asia. Compared to Thailand or Japan, which began their modernization process in the late nineteenth century, these countries are rapidly moving through the steps outlined in table 1.1. An important factor here is the resource-hungry Chinese economy, whose continuous expansion has led to a sharp increase in the value of mining and development concessions.

In Cambodia, it is not uncommon for power-seeking politicians to grant concessions on forests and similar resources to senior bureaucrats (Global Witness 2009). When companies from China, Thailand, and Vietnam offer direct investments to the rural areas of Laos and Cambodia, there is more at stake than merely funding mining operations and other resource extraction activities. The situation also encourages a shift in land use from small-scale farming to large-scale cultivation of cash crops like rubber and sugar cane. This transition can occur very quickly. According to a 2012 study by the University of Bern in Switzerland, foreign direct investment in land rose fifty-fold over the preceding ten years alone, with about 5 percent of the country's territory commodified in this manner (Schönweger et al. 2012). Governments of developing countries are selling large tracts of land through a concession system, using licenses and contracts to grant businesses exclusive rights to exploit specific areas and develop them economically.

As we have seen, governments have gradually expanded their influence over people through environmental policies. However, this may change once they begin implementing significant measures to mitigate the looming effects of climate change. Extreme weather patterns are already driving up food prices in the Middle East. As the Arctic ice cap melts, Canada, the United States, and Russia are becoming locked in a struggle for submarine resources. Competition in technological development is also mounting, spurred by regulations on CO_2 emissions. Confrontations over energy are increasing. For instance, when the Chinese government poured massive subsidies into the solar industry, the United States imposed retaliatory tariffs (Busby 2018). Needless to say, this trade war has intensified since the start of the second term of the Trump administration in 2025. Moreover, countries are facing disasters more frequently. Japan, for example, is building towering seawalls along its northeastern coast to mitigate the effects of potential tsunamis. This is a drastic measure, even for the Japanese, who are accustomed to natural disasters. Updating infrastructure to adapt to climate change is tremendously costly and will permanently alter coastlines and fishing ports as well as the way people themselves relate to the ocean.

1.2.2 When the State Clashes with Local People

Not all states follow the same procedure when converting nature into resources. However, some generalizations about the process can be made. Figure 1.1 provides a conceptual illustration of how the state expands its domains of governance, incorporating one domain of resources after another.[4] As chapter 4 indicates, citizens under the dominance of the state can still maintain a measure of autonomy within the official system. They may, for example, be legally able to establish communal lands.

Frequent clashes and confrontations occur along the borders where the spheres of control of the people collide with those controlled by the state. The fiercest conflicts often revolve around access to land and resources that local people have long relied on for their livelihoods. For example, during the Meiji era, Japan's mining industry intensified, resulting in near-constant clashes between companies, authorities, and local residents over issues like forest access,

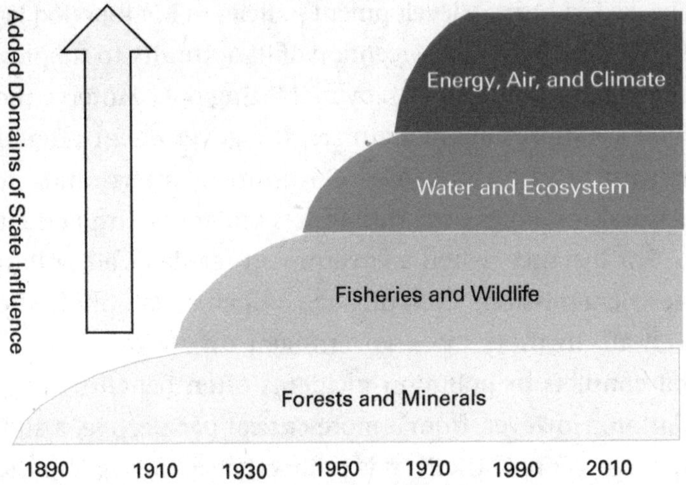

Figure 1.1
The expansion of the environmental state. *Source*: Created by author.

water usage, and pollution. One notorious example is the so-called Ashio copper mine incident in the late nineteenth century, one of the earliest cases of reported environmental damage. Pollution and deforestation related to mining operations caused tremendous environmental devastation in Japan's central Tochigi Prefecture. Around the same time as Ashio, the livelihoods of residents in the town of Kamaishi (in Iwate Prefecture on the eastern coast) also came under threat.

Fishing in the Owatari River, a vital aspect of the local economy, was jeopardized when a nearby foundry siphoned off large volumes of water for its smelting operations. Angry fisherfolk shut off the factory's waterway, leading the company to appeal to the courts for damages (Hayasaka 2013). The central issue was whether the foundry, in its steelmaking processes, had an obligation to ensure that local fishing activities could continue unhindered. Ultimately, the court ruled against the fishing community. This conflict exemplifies the frequent clashes over resource use between residents and state-backed industrial development policies in Meiji-period Japan. It also highlights the state's perception of its authority to suppress local objections to its policies, often by mobilizing police interventions.

When a nation-state modernizes, the government often collaborates with private companies. Government intervention to protect natural resources does not always violate or suspend citizens' rights. For instance, when a government regulates air pollution, it may restrict emissions from private companies to uphold the right to clean air. In these cases, government involvement in environmental conflicts or pollution incidents often benefits the general population. However, from a more critical perspective, a state that enforces emission restrictions is indirectly expanding its influence, encroaching on citizens' rights to public space, and concentrating more authority in central government. It is important to note that such measures may not necessarily reflect a deliberate attempt by the government to extend its sphere of influence.

In some developing countries with significant wealth disparities, the government can conveniently scapegoat ethnic minorities and impoverished citizens. By blaming environmental destruction on individuals who refuse to comply with regulations or possess "strange" customs, the government can increase its control over resources under the pretext of conservation.

When states establish national parks and wildlife sanctuaries, for example, they effectively bring extensive areas of land under state control. Parks were first established in North America and Australia in the nineteenth century, with countries in Europe and South America following suit in the early twentieth century. After World War II, African countries began rapidly establishing national parks (see figure 1.2).

Since their inception, national parks have been sites of conflict. Reports of persecution of ethnic minorities, illegal logging, and displacement of local communities from reserve areas are common globally (Brockington and Igoe 2006).[5] A global study covering 225

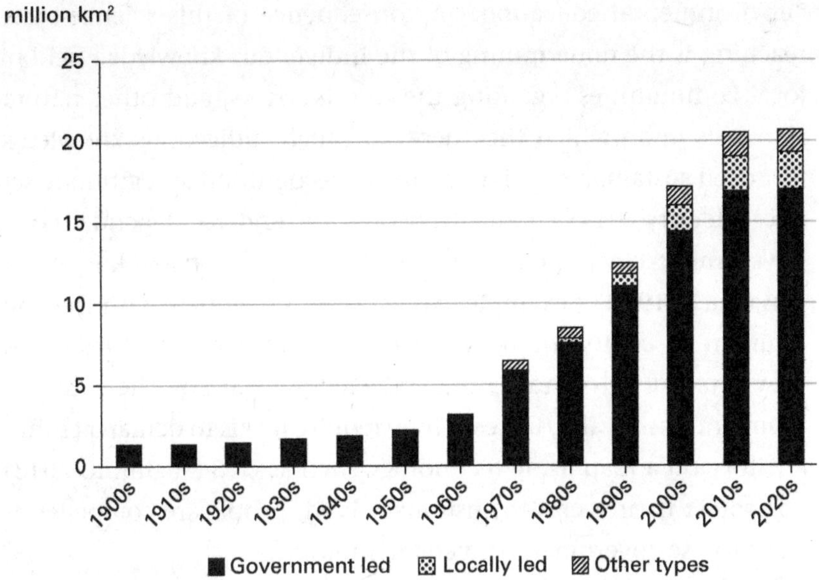

Figure 1.2

The increasing areas of government-led protected areas. *Source*: Based on data from the World Data on Protected areas (www.protectedplanet.net).

protected areas (Zafra-Calvo et al. 2019) reveals that more strictly managed protected areas perform worse regarding local people's access to mechanisms for addressing conflicts and achieving justice. In areas experiencing heavy population growth, people evicted from newly established national parks often clash with original inhabitants in relocation areas.

Over time, governments have refined the techniques used to intervene in nature. They shifted their focus from location-specific resources, such as forests and land, to circulating elements like water and air, expanding control from land to resources. To address issues such as shrinking biodiversity, transboundary pollution, and climate change—problems that indirectly impact the daily lives of many people—governments have begun prioritizing public education. This includes awareness projects about global warming and CO_2 emissions targeting rural populations, often labeled as "environmental education." A consequence of this science-based teaching is the undermining of the Indigenous knowledge held by local communities regarding the forests, rivers, and other natural resources upon which they depend. Much Indigenous knowledge is indeed sustainable and deserving of recognition as legitimate science (Hendry 2014). Gradually, people are rendered dependent on a government-sanctioned system of formal, science-based knowledge (Sasaoka 2019). For example, when citizens negotiate with the government to clarify the boundaries of shared woodland resources, they often dismiss Indigenous knowledge regarding the extent of their lands as invalid. Instead, they require locals to demarcate their territory on a map using technology such as GPS (Sugimoto 2013). To protect their access to customary land, people are compelled to learn to use government-driven technology.

As demands for land grow in developing countries, the question of knowledge invites even more serious consequences. For example, "land grabbing" in Southeast Asia has significantly increased in recent years, at the cost of local residents (Schoenberger 2017). In the early 2000s, the concept of land grab acquired a new dimension: the "green grab," wherein land previously used for food and fuel production is appropriated for conservation purposes (Fairhead et al. 2012). The expansion of palm oil plantations, for example, is often promoted as a crucial step toward increasing the use of carbon-neutral fuels.

The causes, conditions, and consequences of land grabs and climate change politics are inherently intertwined. Effective climate change mitigation and adaptation require addressing global land grabs, and vice versa (Franco and Borras 2019). A case study on Cambodia by Scheidel and Work (2018) demonstrates how norms and beliefs surrounding climate change and forestry policies can further displace local communities from their land and forest resources. Evidence of green grabs highlights the need to pay attention to how global environmental agendas and conversations can result in land use outcomes that disadvantage customary land users.

1.3 Mechanisms of Domination

1.3.1 Wittfogel and Total Power

When a government and its people clash over resources or the environment, we might assume that the conflict concerns only the specific issue at hand. However, this is rarely the case. When state power is concentrated around decisions regarding the use of natural resources in specific sectors, such as fisheries and forestry, the impacts of government interventions often extend beyond the boundaries of these sectors.

German sinologist Karl Wittfogel was one of the first to argue that a country's system for managing resources forms the basis of its social organization. Wittfogel published his argument in the 1957 book *Oriental Despotism*, subtitled *A Comparative Study of Total Power*. He analyzed flood control systems in Asian countries, outlining how cooperation between the state and citizens on water management projects can ultimately lead to the state wielding unfettered and ruthless power (Wittfogel 1957). His analysis was the first clear articulation of how a mechanism for controlling resources can result in state dominance over people.

Let us take a closer look at this pioneering work. Wittfogel began by pointing out that a country's social order is related to how it manages the division of labor, building on and challenging Adam Smith's analysis of this topic in *The Wealth of Nations*. Smith (1976 [1776]) argued that agriculture tends to develop at a slower pace than industry because agricultural work is less suited to the division of labor compared to that in industry. Was Smith correct to characterize agriculture as activity ill-suited to the division of labor? Wittfogel argues no. In his discussion, Wittfogel did not focus on agriculture itself but on the systems of water management that are crucial for successful agricultural irrigation. He noted that these systems require significant division of labor for construction, operation, and maintenance—both to implement irrigation systems and to prevent flooding. According to Wittfogel, these "hydraulic societies" require extensive civil engineering work divided into separate tasks.

Large-scale projects, such as irrigation works, rely on advanced knowledge. Managers of hydraulic economies must possess a sophisticated understanding of astronomy to create calendars and complex mathematical skills to calculate the number of workers needed and maintain records. As Wittfogel stated, "A large quantity of water can be channeled and kept within bounds only by the use of mass labor; and this mass labor must be coordinated, disciplined, and led. . . . The masters of hydraulic society were great builders because they were great organizers; and they were great organizers because they were great record keepers" (Wittfogel 1957, 18).

The intrinsic nature of large-scale irrigation impacts societies in ways that extend beyond their water management practices. Effective resource management necessitates the management of the people engaged in these projects. Those conscripted by the state for massive engineering endeavors can also be mobilized in times of war, illustrating another example of a significant nationwide project. In summary, hydraulic societies depend on an integrated power structure rooted in flood control projects. Wittfogel faced criticism for advocating oversimplified environmental determinism, downplaying human agency in shaping and adapting to the environment (Adas 2001). Nevertheless, his insights provide one of the earliest systematic explanations of how state dominance over nature can translate into dominance over people.

Scholars continue to examine the political and social circumstances of a country in relation to its resource management. Historian Judd Kinzley, for instance, studied the history of the Xinjiang Uygur Autonomous Region, which has experienced worsening ethnic strife since the 1990s in China. Tracing the conflict back to resource development that began in the early twentieth century, Kinzley (2018) argued that parts of the region underwent successive waves of exploration and infrastructure-building, ranging from mining facilities to railroads and other forms of transportation. The outcome has been a widening gap between resource-rich and resource-poor areas. In Kinzley's view, the unmistakable ethnic violence in Xinjiang is symptomatic of the inequalities created by resource development. Once established, economic foundations create favorable conditions for attracting even more investment, increasing profit opportunities, and ultimately becoming sources of political power. As Wittfogel noted, state interventions in resources extend across a country by mobilizing labor from different regions, with the level of penetration increasing over time as investment accumulates.

1.3.2 Popular Resistance

If resource development brings burdens and opportunities to people in specific regions, others will inevitably be left behind. Disadvantaged groups that are excluded will find it very difficult to change their circumstances. However, some options for resistance do remain. When a state generates wealth and power from resource networks, and these benefits become essential for industry, disrupting these networks can be a vital means for challenging the system.

How does this play out in practice? Political scientist Timothy Mitchell studied the history of energy procurement in Western countries, focusing on how democratic societies are influenced by their dependence on fossil fuels (Mitchell 2011). In the past, people primarily used firewood for energy, as forests provided both scattered and abundant sources of fuel. As coal gained prominence, populations became concentrated in specific geographic areas where this new energy source could be efficiently extracted, which, in turn, required a spatial concentration of labor.

How did these population centers foster democracy? According to Mitchell, the Industrial Revolution relied entirely on coal for energy. This dependence empowered coal workers with significant political leverage, from strikes to acts of sabotage, to express their grievances. The new industrial structure enabled workers to make their voices heard, and whenever capitalists and the ruling class felt their income threatened by worker action, they were compelled to listen. Modern industry remains an interconnected web where the failure of one node can paralyze the entire production process. As one leader of the French railwaymen's union explained in 1895, "With two pennies' worth of a certain substance, used in the right way, we can make a locomotive unable to work" (Mitchell 2011, 23).

Due to technological change, eventually, coal was replaced by oil. The industry's shift from coal to oil represented more than just an energy transition; it also meant that workers lost much of the leverage they had when the industry relied on coal. Why? Oil is significantly easier and cheaper to transport than coal. Once energy could be efficiently moved to wherever production costs were lowest, manufacturers were no longer tied to specific energy-producing areas. Management became less reliant on any particular group of workers. Furthermore, while coal mining occurs deep underground, oil extraction takes place above ground and is more automated, resulting in a significantly reduced workforce that is easier to supervise than those working hidden in dark mine tunnels.

In a comparison of the various working environments required to produce and utilize wood, coal, and oil, Mitchell concluded that coal was the most likely energy source to foster conditions for effective resistance by resource workers at that time. Coal miners achieved significant success in industrial action largely because their opponents were economic elites who were utterly dependent on their labor.

1.3.3 Can Government Dampen Development?

Labor movements can utilize strikes and other forms of industrial action to resist developmental states from the bottom up. However, they are not the only stakeholders capable of mounting opposition. Certain groups within the government can also hinder the progress of economic development. As previously noted, countries worldwide have been establishing government agencies for environmental conservation since the 1970s. These agencies can curb rampant overdevelopment and are expected to provide oversight—such as through an environmental assessment system—to ensure that the government does not launch politically motivated development projects arbitrarily.

Despite their legitimate roles and statuses, these governmental agencies face significant challenges. For one, conservation does not yield immediate wealth, while development promises instant financial gain. Therefore, it is not surprising that government agencies advocating for environmental protection often find themselves at a disadvantage compared to those supporting economic growth. Renewable energy development, such as solar and wind, is one area where Asian governments have begun to seek profits as a byproduct of the environmental state (Thurbon et al. 2023). Ultimately, developing nations are eager to catch up and elevate their economies, prioritizing the improvement of living standards over environmental protection. Enhancing the energy efficiency of individual techniques, for example, does not reduce total CO_2 production because they do not necessarily match and cancel out the increasing demand. The marginal position assigned to such agencies will continue to downplay the conservation agenda in growth-hungry countries.

A related challenge arises when environmental protection systems are imposed upon administrative and legal structures designed for production and development. Regardless of how well-intended they may be, environmental protection systems often fall short of initial expectations. Figure 1.3 provides an overview of all ministries and government offices in Thailand involved in environmental protection. The top row shows departments dedicated to production, including agriculture, forestry, and industry. Established in the late nineteenth and early twentieth centuries, these departments have had over a century to develop extensive webs of vested interests. In contrast, the departments within the Ministry of Natural Resources and Environment—focused on conservation and shown in the

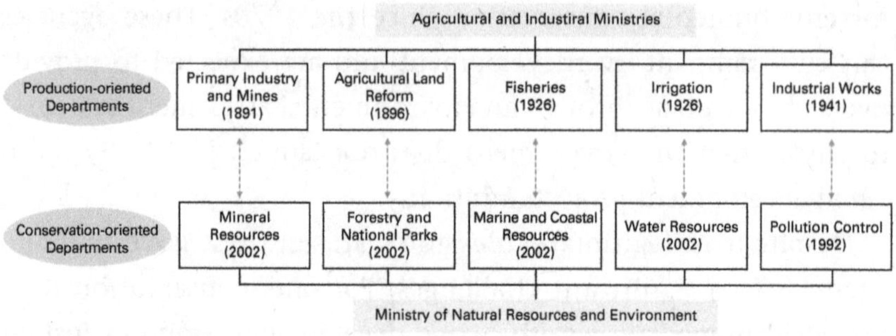

Figure 1.3
Comparison of production- and conservation-oriented government departments in Thailand. *Source*: Created by author.

bottom row—were all established as recently as the 1990s. These departments were structured to function in line with existing laws, without encroaching on the jurisdictions of other administrative agencies, which curtailed their authority from the outset. For example, the Pollution Control Department established in 1992 is authorized to monitor pollutant emissions but can only do so externally, from outside the factories. Its production-oriented counterpart, the Department of Industrial Works established in 1941 on the other hand, possesses the authority to step into factories and grant (or revoke) operating permits (Sato 2013a). It is clear that production-oriented departments have stronger and more penetrating power.

If our administrative system is structurally weak in promoting sustainability, what about the power of the people? Today, states seeking to expand their power in democratic countries rarely resort to overt violations of private property rights or forcible requisitions. Instead, they promote knowledge that supports specific policies, disseminating it throughout society to consolidate a system of rights that reflects those policies. As we discuss in chapter 5, complex, large-scale issues such as environmental degradation and climate change cannot be viewed merely as "problems" without scientific knowledge to back up these assertions. This knowledge must be effectively communicated to the public to facilitate implementation and behavioral change. Consequently, knowledge and information become a political space where the state exercises its power over the people.

1.3.4 Mechanisms of Security

Environmental states employ similar measures to extend their influence throughout society. Comparable to what French philosopher Michel Foucault referred to as "mechanisms of security," these measures arise when a government attempts to eliminate socially harmful incidents such as crime or illness. They involve actions to "normalize" conditions considered necessary for the government's efforts to eliminate undesirable elements. The resulting circumstances are then labeled as standard and acceptable. Here, the government's role is "to establish an average considered optimal, on the one hand and, on the other, a bandwidth of the acceptable that must not be exceeded. In this way, a completely different distribution of things and mechanisms takes shape" (Foucault 2009, 6).

Consider, for a moment, how a government attempts to control the use of limited resources within a region. It may impose regulations on activities such as logging and mineral extraction. In these cases, the intent of the government is clear. However, the situation becomes more complicated when the state seeks to influence complex phenomena like biodiversity or the atmosphere. In such instances, it cannot rely solely on a simple logic of exclusion exemplified by slogans like "Don't steal, don't cut down trees." Likewise, policymakers, no matter how dedicated they are to maintaining public order, can never completely eliminate societal ills. Consequently, a government must determine the level of environmental damage it is willing to tolerate as a result of human exploitation of natural resources. While this level can never be zero, it should not be excessively high either. Crime prevention operates according to similar principles: Statistically, a certain number of people will always violate the law, and the state must account for this reality when determining the acceptable level of dangerous behavior.

In deciding what level of unlawfulness is tolerable, a government considers not only the inconvenience posed by certain behaviors but also the cost of maintaining crime at an "acceptable" level. In other words, the government shifts its focus from managing individual behavior to stabilizing the overall system, calibrating it to ensure that resource use remains within boundaries deemed tolerable. It also establishes "tolerance" levels that society is encouraged to accept.

Consider how this approach applies to natural resource management. When a government establishes a national park in an area populated by settlements, it acknowledges that some local residents will ignore the newly imposed restrictions because their livelihoods depend on them. The government then accommodates these residents, at least to some extent, creating a system that functions much like a "mechanism of security." Pollution problems often unfold in a similar manner. It may not always be feasible to eliminate or drastically reduce harmful emissions. Considering available technology and the incentives motivating all stakeholders, the challenge lies in finding the level of pollution that is tolerable for both those responsible for the emissions and those affected by them.

Foucault identified the emergence of a security mechanism at work when vaccination was employed to eradicate smallpox in nineteenth-century Europe. Smallpox vaccinations involved inoculating individuals with a small number of pathogens from a similar poxvirus, which produced antibodies to protect them from the more deadly smallpox variants. This practice of artificially triggering a less severe form of illness to prevent a more serious affliction represented a radical departure from the prevailing medical thinking of the time. The effectiveness of the vaccination derived from the skillful manipulation of what constituted an "acceptable" level of illness, which, for Foucault, exemplified a typical mechanism of security (Foucault 2009, 74). Once a disease is controlled at the individual level, the government can estimate infection and mortality rates based on specific demographic factors. This allows the state to approach risk management as a matter of applying science and statistics. At this point, the determination of "acceptable" levels falls to bureaucrats and state-supported scientists.[6]

The environmental state is a phenomenon that reflects how the state's environmental interventions alter relationships among people. Over time, the environmental state also becomes the gatekeeper of the norm, embodying the particular interests of those in power. Governments and scientists work to normalize a specific relationship between humanity and nature, encouraging citizens to embrace that norm. Questions arise: How much of a country's territory should be forested? What about national parks? To what degree of energy independence should a country aspire? How much pollution should it tolerate? While these goals may appear neutral, their practical implications are far from it.

On the surface, the security mechanisms employed by an environmental state often appear moderate and attuned to the circumstances of the people within that natural environment. However, it is crucial to remember that governments hold exclusive authority to determine what levels of inconvenience are tolerable and the consequences for those who exceed these levels. For instance, successive forest conservation policies throughout Southeast Asia have normalized the idea that national parks must be protected. If stringent protection is deemed the standard, then locals practicing slash-and-burn agriculture in or near national parks are often labeled either antisocial or incompetent. Consequently, governments and the international community face a pressing dilemma: how to discipline or educate these "uncivilized" individuals to ensure compliance with the established norm. The environmental state then attempts to win hearts and minds through "environmental education," a type of government intervention that appears kind and benevolent.

We may conclude that the character of a particular environmental state is shaped by how citizens react to the levels and ranges of tolerance established by government policy. At times, individuals may oppose government policy, while at other times, they comply with little complaint. Those who experienced the 2011 Great East Japan earthquake continue to grapple with government-determined tolerances. Following the Fukushima Daiichi nuclear plant accident, many concerned citizens began voluntarily measuring radiation levels in their neighborhoods, generating a wealth of data on contamination levels throughout the country. However, it is ultimately the government that sets the standards for what levels of pollution are considered harmful to human health, and the state must delineate these boundaries. While citizens armed with Geiger counters can measure radiation levels, this ability does not empower them to accurately define what is acceptable.

1.4 Nature as a Channel for Control

Conventional discussions of inequality and justice tend to focus on how societies distribute financial resources and political power. Today, the environment has become a new domain where the state can establish rules and expand its powers, leading to significant implications for social inequality. Government policy regarding nature is closely intertwined with the livelihoods of populations, particularly those who directly depend on it. The state controls many key sectors, including disaster response, water, and energy supply. In essence, an environmental state impacts the very foundation of human life, making the distribution of natural resources and the responsibilities of environmental protection a matter of justice as well.

It took considerable time for public discourse on environmental policies to consider their justice dimensions. For decades, developmental states viewed nature primarily as a source of raw materials and a site for waste disposal. By the time the environment became a focal point for state intervention, it had largely gone unnoticed by the public. At that juncture, technological development had already alienated urban residents from nature, making it difficult for them to perceive changes in the environment. Many individuals struggled to understand why environmental justice should matter to them. It is easier to discuss government policies when people can see how these might provide them with tangible benefits—whether in terms of income, rights, burdens, or something else. However, environmental policies are rarely so straightforward.

Today, justice and fairness have taken center stage in debates about how people relate to one another and the environment. The water, air, soil, and forests that governments regulate, at least on paper, are shared resources for all. Yet, the rights and burdens associated with these resources are not distributed equally. Some individuals use—or pollute—more than others, and not everyone is compelled to make sacrifices for a cleaner environment. While abundant land and water might not initially seem like a justice issue, the reality is that certain resources—such as arable land and clean water—are finite and thus are inherently political. This scarcity ignites conflict among individuals, profit-seeking businesses, state interventions, and the problem-solving efforts of international organizations.

In his 1983 book *Spheres of Justice*, political theorist Michael Walzer argued that structures of control function through what he calls "social goods," which are items that can be distributed and managed among people. Unlike an individual's innate capabilities—which cannot be divided or shared—a social good is not inherently tied to any particular person. The distribution of social goods necessitates cooperation. In Walzer's words, "The means of domination are differently constituted in different societies. Birth and blood, landed wealth, capital, education, divine grace, and state power—all these have served at one time or another to enable some people to dominate others. Domination is always mediated by some set of social goods" (Walzer 1983, xiii).

What happens when the natural environment becomes a part of the social goods that mediate domination? The governance of the natural environment is further complicated by the fact that it can confer rights or social goods to individuals, which are alarmingly easy to convert into other forms of advantage. An individual who gains governance over this sphere can leverage these tools to extend their control. For instance, a person who dominates through substantial land ownership is likely to find it relatively straightforward to translate that power into political and economic influence. Similarly, a skilled politician or bureaucrat may naturally gain influence in politics but could choose to extend their reach into business through dubious methods, potentially becoming an unstoppable tyrant. The consequences of specific distribution choices on the environment are difficult to predict. Therefore, it is crucial to scrutinize these choices and investigate the biases that can lead to significant injustice.

Over the years, different societies have held varying ideas about what constitutes "justice." Nevertheless, a consensus has emerged across time and geography regarding what defines extreme *in*justice (Sen 2010). Genocide, human rights abuses, structural violence, and other extreme injustices do not occur simply because a central authority desires these atrocities. Persecution does not happen without the support of the masses who form the state's power base.

Developmental states have pursued modernization and economic growth at all costs, often downplaying the environmental destruction caused by their large-scale consumption of resources. The riches promised by development were so alluring that they overshadowed any consideration of negative externalities. This developmental mindset has persisted in the renewable energy sector under the guise of "green" innovation. It is tempting to believe that countries transform into environmental states after reflecting on their reckless economic expansion. However, the reality is that environmental states seldom pause to consider how their conservationist zeal can impact human societies. It is no surprise that these impacts often go unnoticed. The increasing regulation of people is an overlooked side effect of state-led conservation, even when it becomes far-reaching. A contributing factor to this invisibility is that those disadvantaged by new policies are often already politically marginalized.

Today, state intervention in the natural environment and the accompanying injustices have become so significant that they can no longer be ignored. When states advocate for economic development while tolerating poverty and inequality, they risk revolution and social unrest. The same principle applies to environmental policies: If these policies are not grounded in justice, they cannot sustain ecosystems and may consequently jeopardize human life.

Anyone analyzing interpersonal relations in the context of the environment must keep in mind that the state always plays a crucial role, either overtly or otherwise. State-led development that converts nature into a resource has tangible consequences for the environment. This progression is followed by state-led policies aimed at protecting that same environment, which in turn reinforces the process of turning nature into politically useful resources. Along the way, the state consolidates its power further. The appropriation of common resources that once represented particular areas—such as shared forests or grazing lands—is suddenly legitimized by official, feel-good designations like "community forests" or "community irrigation infrastructure." State intervention and approval are required for any action that affects the environment, whether it involves resource development, conservation, or determining which of these two priorities takes precedence.

Environmental policies have evolved beyond their original goal of managing the natural environment. They are now emerging as forces that seek to modify nature in ways that alienate local residents while enhancing their power as environmental states. These states influence all aspects of human society, and we must consider the implications of this shift. In the past, it was sufficient to implement targeted environmental policies aimed at protecting specific elements of nature, such as forests or water resources. That era has passed. In the face of global climate change, we need policies that recognize humans as an integral part of the environment.

2

Knowledge: A Scientist Stands with Pollution Victims

The tremendous power of the state is not always the determinant of inversion; rather, inversion is often facilitated by the local people who are forced into accepting the state's logic, which is often backed by science. How are different types of knowledge related to the rise of inversion? How can the experiential knowledge of the victims inform policies that tend to downplay and neglect their claims? In this chapter, I focus on the role of a unique Japanese scientist, Ui Jun, who organized an informal course called *Kōgai Genron* (Principles of Pollution) in the 1970s at the University of Tokyo. Striving to challenge the polluting companies and their supporting scientists, Ui's analysis provides insights into how the hierarchy of environmental knowledge can lead to inversion.

2.1 State Inaction in Environmental Problems

Japan has long experienced severe environmental problems, especially in the 1960s and 1970s. Environmental pollution issues—represented by Minamata disease, Itai-itai disease (water contamination due to mining), and Yokkaichi asthma (pollution caused by sulfur dioxide)—have all become well-known phenomena posing serious threats to life and health to local population. These problems were so severe that they prompted the creation of a new word in Japanese: *Kōgai* (literally meaning the destruction of the public domain) (Ui, 1992).

Minamata, the name of a city that later became the name of a disease, is one of the most notorious of the four major pollution cases that occurred in Japan from the 1950s to the 1970s due to industrial wastewater pollution.[1] It broke out in a fishing town in Kumamoto Prefecture, a southern region of Japan. First identified in 1956, it took nine years for the government to officially acknowledge that it was caused by mercury-contaminated water emissions from a factory in Minamata owned by the petrochemical corporation Chisso. Even today, despite many years of effort from countless individuals and organizations, reliable data on how many people have perished due to Minamata disease remains unavailable.

It was one of the first cases of industrial pollution to attract global attention, largely due to the shocking photographs of pollution victims taken by American photographer Eugene Smith, published in *Life* magazine on June 2, 1972. The "strange disease" first appeared to take on epidemic proportions in early 1956 and ultimately became known as "Minamata disease" (Smith and Smith 1975, 28). After a series of victims were hospitalized with similar symptoms of brain damage, a group from Kumamoto University reported in 1959 that organic mercury was the cause. In October of that year, the Fishermen's Union stormed the factory, demanding further indemnity and a cleanup of the bay. Ultimately, the fishermen settled for meager payments after facing repeated threats from the company and warnings from the government to accept the compensation offer.

However, confusion about the real cause still prevailed, with many suspecting that Chisso, a local petrochemical and plastics manufacturer, paid experts to refute any report that might damage their position (Smith and Smith 1975, 30). In reality, between 1932 and 1968,

Chisso Corporation dumped an estimated twenty-seven tons of mercury into Minamata Bay, poisoning fish and, eventually, the people who consumed them.

Even more remarkable than the severity of the pollution was the way the sociopolitical system further exacerbated the damage. The risk of water pollution was recognized as a potential threat to health and local livelihoods in Japan as early as the 1890s when the notorious Ashio copper mine pollution attracted public attention.[2] Other smaller-scale pollution incidents have been reported intermittently as Japan's industrialization progressed (Terao 2009).

Minamata is not just a disease of the past. In 2004, more than sixty years after its initial discovery, the Supreme Court ruled against the national and Kumamoto prefectural governments for negligence in allowing the spread of mercury poisoning near the Shiranui Sea in western Japan (Yoshinaga and McCormack 2004). The Japanese Ministry of the Environment reported that out of 35,092 applications, only 2,998 were newly acknowledged during the 2009–2012 special relief scheme established by law (Ministry of the Environment 2020).

The most recent decision regarding Minamata was handed down in September 2023 (*Japan Times* 2023). In this case, the Osaka District Court ordered the central government, Kumamoto prefectural government, and Chisso Corporation to pay damages to plaintiffs from thirteen prefectures who were excluded from a special relief program for those affected by Minamata disease. While the number of medical cases has naturally decreased due to aging, the institutional battle continues to this day (see figure 2.1). The sharp decline in applications after the mid-1990s is attributed to a political resolution in 1995, wherein the government and companies agreed to provide lump-sum compensation, which discouraged some individuals from seeking official recognition. However, following the Supreme Court decision to relax eligibility criteria, the number of applicants increased once again.

The rise and fall of applications depicted in figure 2.1 does not signify the changes in the actual number of victims. The number

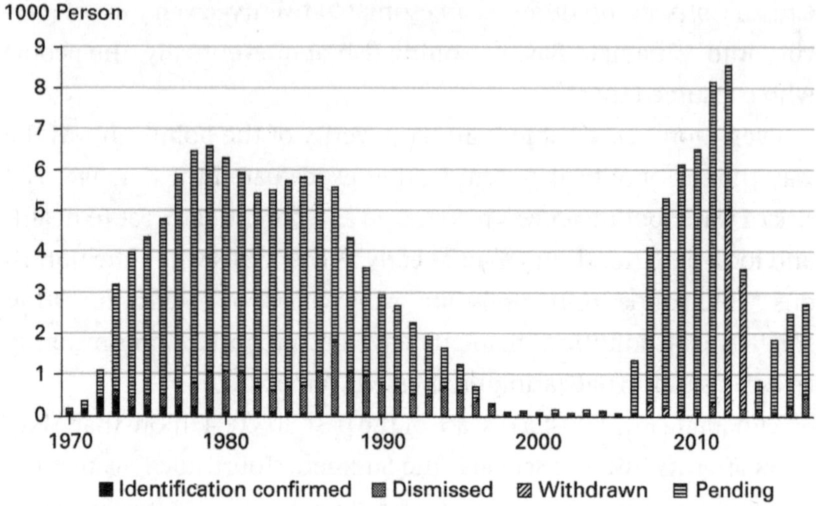

Figure 2.1
Applications for Minamata disease victim identification (annual). *Sources*: Before 2000: calculated by the author based on data from Kumamoto, Kagoshima, and Niigata; 2001–2020: Ministry of the Environment of Japan (data for 2018–2019 are not available), the Supporting Center for Minamata Disease, the Government of Kagoshima Prefecture, and the Government of Niigata Prefecture.

is more of a function of state-defined eligibility criteria defined for who can be supported.

In the preface of this book, I defined "inversion" as a process initiated by an environmental policy in which potential local collaborators become adversaries, thereby damaging the very objectives of that policy. Although compensation for Minamata victims is not an environmental policy in the strictest sense, it still represents an effort to mitigate the harm caused by environmental pollution. The broader implications of this policy are worth discussing, as they may impact the future behavior of individuals facing similar pollution cases.

"Money has invaded the villages of Minamata," wrote Eugine Smith (Smith and Smith 1975, 145). In 1973, after a series of court trials, the Central Pollution Board decided to pay $68,000 for severe cases and $60,000 for less severe cases (Smith and Smith 1975, 145).

Eugine Smith documents a comment from a victim who noted that everyone from bankers to furniture chain stores to home builders "came to this shabby old place asking us to invest our indemnity money," but "some had been so scarred by the ostracism attached to their illness that they dared not use their money openly" (Smith and Smith 1975, 145).

US-trained sociologist Kazuko Tsurumi, known for her work on endogenous as opposed to externally driven development, conducted five years of fieldwork in Minamata in the late 1970s, specifically on this topic. Tsurumi observed the divisions within local communities created not only between the pollution victims and those who escaped but also among the victims who received compensation from the Chisso company and the government. She noted, "Due to the pollution of the Shiranui Sea, fishermen can no longer live there without compensation, either for fisheries or as certified Minamata disease victims. However, the indemnity money created conflicts among the local people" (Tsurumi 1998, author's translation). Within the same neighborhood, some individuals received compensation while others did not. Over time, the large indemnities began to pacify the momentum of local fishermen's protests; however, this silencing was not solely due to compensation. Rather, it stemmed from self-restraint among the local people and the divisions within the community, which made it more challenging for them to unite.

Tsurumi's work further demonstrates that conflicts have arisen not only between those officially certified and those who were not, but also between those certified early and those certified late. Materially, this conflict manifested in the survival of individuals who became reliant on support provided by the state. The community that once functioned as a social buffer against external shocks is now destroyed—less by the pollution itself than by the manner in which indemnities were provided. This represents one type of inversion, where policy fails to recognize that human relations can be damaged by pollution as well as by the distribution of funds intended to address the problem.

The failure of various levels of government in the late 1950s and early 1960s may not be surprising. During this period, Japan was an archetypal developmental state. As pollution began to be more broadly recognized as a social issue, Japan was experiencing rapid economic growth later termed a "miracle" by the world media. At that time, there was a strong tendency to accept pollution as an "inevitable strain of rapid economic growth." Inversions occurred when the government sought to resolve environmental problems through monetary compensation, an economic solution that left the fundamental power structure unchanged. Below, we will explore the movements of outsiders committed to Minamata and other pollution cases as an opportunity to change the very power structures that invite such inversions. This case will illustrate how inversion is not merely a "local" phenomenon, with local actors taking the initiative, but also how concerned outsiders can help address the issue by connecting local struggles to a broader structure of political economy dominated by governments and large corporations.

2.2 *Kōgai Genron* (Principles of Pollution)

2.2.1 Ui's Rebellion

While a nation-state rapidly modernizes, the government often colludes with larger private companies, which are the main drivers of economic growth. Fighting against pollution, therefore, means challenging this power structure; it does not merely address pollution as an unintended consequence of state-led development.[3]

It was Ui Jun, a vocal researcher (see figures 2.2 and 2.3), who pointed out that pollution was not an unintended consequence of economic growth but rather a systematic part of it. Starting his academic career as a water engineer, Ui sought to redefine the Minamata problem from the perspective of its victims. He began to question

Figure 2.2
Voluntary open class to "dismantle Tokyo University" (ID: P100413035644). *Source:* Asahi Newspaper Photo Archive.

the role that outside researchers and universities had played in the pollution agenda. According to Ui, these external researchers not only legitimized prevailing beliefs but also actively mystified the true causes of pollution. Observing that industry representatives were rationalizing pollution as an unavoidable side effect of economic development, Ui countered that "ignoring pollution was a factor that allowed rapid economic growth" (Ui 1971, 225). By neglecting the costs of environmental damage, Ui argued, private companies could profitably prioritize their investments in production. This oversight, coupled with the use of underpaid labor, was, according to Ui, a significant driver of economic growth in Japan (Ui 1992).

Although based at the prestigious University of Tokyo, Ui remained an "assistant" (*joshu*, the lowest academic rank) for more than twenty years. Not entitled to offer official courses, he began an

Figure 2.3
The final lecture of Assistant Professor Jun Ui in 1986 was a great success with a full house (P151231004314). *Source*: Asahi Newspaper Photo Archive.

unofficial "voluntary course" in the evenings open to the general public. This initiative garnered him fame despite his modest status in the otherwise privileged Faculty of Engineering at the top university. Titled *Kōgai Genron* (Principles of Pollution), the lecture series started in 1970 and continued for fifteen years, attracting over a thousand participants at times, including student leaders, activists, and actual victims of pollution. This group became a driving force behind the subsequent antipollution movement in Japan, serving as intermediaries to the state, private companies, media, and most importantly, the pollution victims. From the perspective of the senior professors and executives at the University of Tokyo, whom Ui had criticized as "a group of scholars serving the governmental authorities," his voluntary lectures must have felt like a guerrilla-style rebellion. His opening remarks at the first session illustrate his stance on power and hierarchy within the scientific community:

> In the case of pollution, universities and colleges are significant complicit actors. In nearly every pollution incident, the university and its graduates contribute to worsening the situation. A typical example is the University of Tokyo. When we consider this university as an illustration of how a national institution—supported by state funding and infrastructure and established to educate individuals for national service—has become an instrument of oppression and discrimination against those who uphold the state, we see a stark contrast with the University of Warsaw. The latter was secretly and continuously developed as a base of resistance. Studying there involved risking one's life without any associated privileges. The ultimate goal of studying was the necessity for survival, unrelated to career advancement, which encapsulates the essence of the course "Principles of Pollution." (Ui 1971, 2, author's translation)

Much has already been written in Japanese about Ui's legacy, and I do not claim the authority to evaluate his achievements either as a water engineer or as an elucidator of Minamata disease.[4] My focus is on the stratification of our knowledge of the environment. As will be elaborated in the next section, Ui repeatedly emphasized that this stratification is the root cause of pollution and the ongoing social damage that persists even after technical treatments are employed to eliminate the pollutants. The core issues of knowledge and power will become increasingly apparent as the state solidifies its dominance as the primary entity "solving" climate and environmental challenges. Furthermore, it is the culture of academia—where certain forms of knowledge are regarded as valuable while others are dismissed as inconsequential—that leads to interventions that, although intended to assist people, too often entrap and constrain them.

2.2.2 The Class System of Knowledge and the Third Party

Ui was a vehement critic of Japanese elite society, particularly regarding how research and education are structured to reinforce the hierarchy of knowledge. This hierarchy is difficult to change, as it relies on "theories thought up by others" based on the existence of "precedents" (Ui 1971). In the Japanese context, he viewed the University of Tokyo as occupying the apex of this system and therefore the primary target of his criticism.

There are three major areas of significances of Principles of Pollution through which Ui sought to challenge the existing hierarchy of knowledge. First, Ui viewed pollution in Japan as an intellectual opportunity for Japanese people to confront a problem that is evident to all and to address it on their own terms rather than through imported intellectual tools. In his view, it represented "a challenge that Japan truly should welcome" (Ui 1971, 232). Ui's endeavor had the potential to fundamentally challenge the prevailing decision-making culture, which often evaded responsibility during disasters by labeling such events as "unprecedented." Japan has experienced frequent disasters, including earthquakes, tsunamis, and typhoons, resulting in substantial damage. Bitter lessons have accumulated about the dangers of relying on "precedent," exemplified by the Fukushima nuclear accident in 2011. The high frequency of unpredictable disasters stimulated a dominant mindset focused on a stable and predictable trajectory of economic development and the associated improvements in human conditions that such development is expected to bring.

Second, Ui describes the antipollution movement as "the story of a losing battle" (Ui 1971, 231), but he argues that the experience of "losing" offers a more beneficial learning opportunity for Japan and other countries. Japan's history of negative experiences with seemingly unsuccessful civil society movements offers a valuable reference point for developing countries today, where the dilemma between top-down development and environmental conservation is as pressing as it was in Japan during its period of rapid economic growth. Civil society's power struggle in these countries might benefit from the Japanese grassroots experience of combating large companies, preparing for legal battles, and mobilizing scientists while preventing ordinary citizens from being exploited.

Third, there is an opportunity to appreciate the tacit yet comprehensive knowledge of victims, which tends to be downplayed by scientific knowledge, as companies and governments often seek to defend themselves. Ui believed that the general public, including pollution victims, possesses experiences that extend beyond what can be captured by scientific knowledge. He characterized the victims' systemic and comprehensive awareness of the pollution experience as "inexpressible."[5] What is crucial here is not just the existence of such tacit knowledge but the arbitrary elevation of one form of knowledge as superior while marginalizing another—a key issue that Ui found problematic.

The conventional framework of benefit and suffering, often expressed in quantitative terms, disregards victims' "inexpressible experiential awareness" (Ui 2000, 51), making it difficult to adequately address their voices. More importantly, this disregard squanders knowledge and skills that could be useful in tackling real-world problems. In today's world, where modern scientific thinking is more accessible through the internet, locally based tacit knowledge seems increasingly downplayed. Between privileged formal knowledge and local tacit knowledge exists a disparity that Ui described as a "class system of knowledge."

Ui's battle with the establishment led him to conclude that there is no such thing as a "third party" to a pollution problem. The third party refers to scientists and researchers who claim to be neutral and to not align with either perpetrators or victims. According to Ui, anyone or any organization claiming to be a third party will inevitably side with the perpetrator, regardless of their intentions to remain neutral.[6] This is because the problem cannot be defined using objective data—instead, it is the implicit framing of the problem that dictates what *becomes* the data. The following observation summarizes Ui's point:

> When damage is perceived, the victim feels it with their whole body. Objectifying such an experience so it can be communicated to others using words is not easy, and in many cases, victims cannot adequately express it. The victims' perception of pollution is thus more systemic and comprehensive. In contrast, the perpetrator's perspective—limited to quantifiable aspects such as the concentration of pollutants or the number of victims—is at best partial. If someone claims to be an impartial third party and tries to hear both sides equally, their perception falls somewhere between partial and total and they will inevitably align with the perpetrator. (Ui 2000, 51, author's translation)

Ui's argument is that while quantitative representations of problems may effectively communicate issues beyond their immediate context, they can never fully capture the totality of the victims' experiences. Conversely, although victims possess intimate knowledge of their experiences, these experiences are difficult to communicate. No matter how comprehensive and rich a personal experience may be, from the perspective of aggregation, each unique experience is, at best, a single data point and, at worst, a noise that obscures the understanding of the broader population. There is a huge task here for anyone interested in knowledge communication. One direction is to make science more relevant to the unique experience of people, and the other is to communicate the value of tacit knowledge to the scientists and policymakers. Victims' experiences should be given legitimate weight when judging the state of affairs.

Ui underscored that the "governance of knowledge" lies in how we understand pollution before taking action to address it. Local knowledge, despite its depth and contextual understanding of the problem, has often been overlooked and silenced in favor of theoretical knowledge, which tends to be applied universally regardless of the specific location or context.[7]

2.3 Governance of Knowledge

2.3.1 How Things Could Be Done Differently

The governance of knowledge—the process by which specific knowledge acquires legitimacy—consolidates the class system of knowledge. Reviewing the history of the twentieth and early twenty-first centuries, we observe that this governance has developed with a bias toward formal scientific knowledge. However, there was a time in ancient Greece when tacit knowledge held equal status with formal knowledge. It is important to consider how tacit knowledge was understood and valued during that period.

In *Nicomachean Ethics*, Aristotle classified knowledge into three categories: *episteme* (true knowledge, science), *techne* (knowledge of craft and art), and *phronesis* (knowledge of doing, ethics, and prudence). Among these, "science" refers to knowledge that can be taught by others and is constant across time and space. While episteme aims for a universal explanation, the other two categories pertain to tacit knowledge, where multiple ways of being are possible without necessitating convergence toward a single best approach. Techne refers to the "craft involved in making things happen," what we now call technology. The other category, focusing on knowledge of action and practice, is what Aristotle termed *phronesis* (Aristotle 1954).[8] While the first two types of knowledge have evolved into what we regard today as science and technology, *phronesis*—often translated as ethics or practical knowledge—has not developed a coherent successor concept (Flyvbjerg 2001) and has remained relatively stagnant.[9]

What are the major characteristics of tacit knowledge? The first is its integrative nature for practical purposes. Formal science divides objects of observation into parts for analytical examination. The results of these analyses typically provide findings on statistical trends and correlations based on selective observations of samples. However, in addressing real-world problems, integration of relevant information is essential for effectiveness. This integration relies on tacit knowledge to intuitively grasp the unique circumstances of a particular place and time. In agriculture, for example, crops grow well only when specific local knowledge about the soil, water, the climate, and the crops themselves is integrated toward the goal of stable cultivation.

The second characteristic of tacit knowledge is that it is learned and acquired through personal experience. Even individuals who can perform everyday tasks—such as cooking or riding a bicycle—or engage in professional practices like agriculture or medicine may struggle to articulate how they complete these activities.

A third characteristic is its focus on context-specific situations and individual cases. While universal formal knowledge tends to overlook context, tacit knowledge produces practical effects *through* its ability to understand and engage with specific circumstances. Because of this, Aristotle argues that sometimes "those who do not possess knowledge are more useful in practice than those who do" (Aristotle 1954, 146). An illustrative example follows: "For if a man knows that light meats are digestible and wholesome but does not know which meats are light, he would not promote health. Conversely, the man who knows that chicken is wholesome is likely to foster health" (Aristotle 1954, 146).

The knowledge that chicken is healthier is indeed specific. Aristotle also discusses a physician who makes mistakes in treating a patient: "If then, one were to have a theoretical account of the immediate particular, knowing the universal but being ignorant of the immediate particular, he will often err in his treatment. For it is the particular that must be treated" (Aristotle 2004, 5). In everyday practice settings, it is crucial to possess the ability to interpret the unique context that shapes the situation.

So far, we have viewed the victims as the primary holders of tacit knowledge, but this perspective does not fully explain the prolonged inaction surrounding Minamata. It was not merely that the victims had waived their right to claim by accepting a settlement; the proprietor's and the general public's sense of discrimination against the victims—particularly the fishermen in rural areas—also contributed to this situation.

Viewing victims solely as objects of pity or as sanctified figures diminishes their political agency, depriving them of the right to express themselves (Hirai 2015, 90). This leads to the disempowerment of tacit knowledge by those in positions of power, exacerbating the divisions among the victims. The marginalization of tacit knowledge thus creates a double-edged sword: on one hand, it hinders less educated individuals from having a legitimate voice in political discussions; on the other, it conceals the unspoken biases of those in power. Together, these dynamics perpetuate the problem without contributing to a fundamental solution.

We know from experience that the knowledge of people living on-site is indispensable for managing local resources such as forests, land, water, and fish.[10] Such knowledge can also act as local sensors to detect changes in the quality of these resources due to pollution or other climate impacts. As the debate on climate change and geo-environmental science escalates, disciplines that capture local specificity are increasingly necessary.

In the Minamata case, scientists attempted to apply their knowledge through an experiment to determine whether Minamata-like symptoms would occur in cats injected with wastewater from the factory. On July 21, 1959, Dr. Hosokawa Hajime, a medical doctor at Chisso Hospital, confirmed that a particular cat labeled "No. 400" exhibited the same symptoms. However, after consulting with factory management, Dr. Hosokawa decided not to disclose this information, arguing that it was "only one sample," which was insufficient, for him, to claim scientific validity (Kobayashi 2021, 73). It was not the science itself that failed to serve the local people.

Setting aside the probable defensive attitudes among Chisso employees, it is essential to acknowledge how science was employed to dismiss concerns about Minamata disease. The small sample size (in this case, $N = 1$) was used as justification to dismiss the results of the experiment. But this begs the question: Why did the experiment stop after testing only one case? If one cat could be tested, surely more cats could have also been tested so that the sample size of the experiment would have satisfied scientific standards with "enough cases" (Matsubara 2002, 191). A critical question is whether appropriate incentives existed to increase the sample size to begin with and whether there was enough time to wait for the testing result to inform policy.[11]

The issue of tacit knowledge and sociopolitical biases not only prompt reflection on the science we have relied upon to understand environmental issues but also leads us to more profound questions about the structure of the state that claims to rely on scientific evidence. Modern science served as the intellectual foundation of the developmental state and has now been adopted as the foundation of the environmental state. If "environmental problems have emerged as a result of the contradictions of modern science" (Torigoe 1989, 18), then it is important to examine the process by which scientific knowledge is privileged within the environmental state.

2.3.2 The Privilege of Scientific Knowledge

Throughout history, political judgments and governance in many countries have relied, in fact, on highly personal tacit knowledge, such as the intuition and connections of a limited number of rulers (Berlin 1996). Conversely, the ruling class often perceives the diverse and complex tacit knowledge among the governed as a potential threat to their power. Even in democratic societies, science has frequently been marginalized in justifying predetermined policy directions. However, when it comes to uncertain issues with significant political implications—such as environmental problems—science becomes politicized and plays a critical role in justifying (or negating) certain courses of action.

What is the problem with the state's approach to knowledge? Political scientist and anthropologist, James C. Scott provided a sharp critique of issues related to "seeing like a state," which inherently tends to marginalize local knowledge (Scott 1998). He described how various simplification procedures shape reality, allowing central governments to operate from a distance. By focusing on how the concept of modernization explicitly transformed the objects of governance, Scott used examples such as the household registration system, language, weights and measures, and property rights to illustrate how both rural and urban spaces became subjects of state control (Scott 1998). As it became possible to gather and manipulate information from across the country from the comfort of offices in the capital, local knowledge was diminished in status and overwritten by "legible" uniformity.[12] In the context of Southeast Asia, slash-and-burn shifting cultivators in remote mountains or slum squatters in urban settings exemplify "illegible" people from the government's perspective. In these instances, various forms of resettlement programs "properly" rendered subjects legible and taxable.

When we examine the conflict through the lenses of both "those who govern" and "those who are governed," we uncover a deeper issue in their relationship than merely the coexistence of different forms of intelligence: the forgetfulness of people. As discussed in the next section, if we forget about an issue entirely, the debate between scientific knowledge and tacit knowledge loses its relevance.

2.4 Stages of Forgetting

Up to this point, the discussion has assumed that knowledge is always present, waiting to be utilized. However, a major factor that complicates environmental issues as a governance challenge is our tendency to neglect the lessons of the past, even when they are readily available.

Everyday environmental risks, such as air pollution in major cities like Beijing, are more readily recognized as "problems" than natural disasters that occur once in a millennium. The less frequently disasters occur, the more people tend to forget their existence and the pain they cause. In the broader context of evolutionary history, the ability to forget traumatic experiences has been a significant strength for humans.[13] This remains true to the extent that individuals cannot influence their social or material environment. However, it becomes a weakness when we fail to apply the lessons learned from past calamities in an age when humans can—especially collectively—shape our environment.

Despite efforts to learn from the experiences related to pollution through textbooks, museums, documentary films, and storytelling for site visitors, the problem of acknowledging the victims of pollution persists.[14] Additionally, new issues of water pollution continue to emerge, including carcinogenic chemicals such as PFAS, which can significantly impact human health (Tang et al. 2023).

Scientific understanding of the warning signs of unexpected disasters has improved, and the accuracy of predicting the time between an earthquake and a tsunami has increased dramatically. However, the risk of natural disasters remains a probability. While science serves as a valuable reminder of past experiences, it can also complicate matters when it comes to attributing responsibilities and compensation in the aftermath of such disasters.

It is not solely the limitations of human memory or a lack of ability to record events that drives past lessons into oblivion; many experts have actively "helped" people to forget. Ui, reflecting on his own experiences with pollution issues, outlines a common pattern in nearly all instances of pollution. The narrative unfolds as follows: When pollution is discovered and its cause is about to be identified, a "neutral third party" emerges, claiming that "the real cause lies elsewhere." This diversion leads to the true cause being obscured, often resulting in the problem itself being forgotten. The key point here is that the forgetting of past lessons does not always stem from poor human memory but is often actively and purposefully encouraged.

In each time period, certain individuals possess foresight and are one step ahead of their time. However, this does not necessarily mean that their findings are reflected in policy or that their conclusions are empirically valid. Historically, the majority of environmental damage is cumulative and has stemmed from a prolonged delay between discovery and action. Still, some have perceived the problem more clearly than others. The question of how to shorten the gap between the recognition of an issue by vulnerable populations and the actions taken by those in power to resolve it is as crucial as developing technologies to mitigate environmental risks. In fact, developing the means to appreciate inexpressible memories may be a much faster path to preventing inversion than waiting for a technical response to problems after they become apparent.

2.5 *Kōgai Genron* and the Environmental State

"The fox knows many things, but the hedgehog knows one big thing." This quote from the ancient Greek poet Archilochus, cited by British political philosopher Isaiah Berlin, categorizes great writers and thinkers into two groups: hedgehogs, who strive to relate everything to overarching theories, and foxes, who draw upon a wide variety of experiences and believe the world cannot be simplified to a single idea (Berlin 1953).

In examining the behavior of the environmental state, it seems that the central government functions as a hedgehog, establishing the policy direction for a multitude of foxes—local individuals who possess a vast range of diverse knowledge based on their specific contexts and interactions. In addressing uncertain environmental events, the government seeks scientific advice and aims to exert full control over the distribution of benefits and burdens. This drive toward centralization, bolstered by uncertainties surrounding these events, grants the state a powerful authority that can become irreversible.

In a world of increasing uncertainty, what form of knowledge is useful in addressing environmental issues? By examining the history of academia's collusion and exploitation by those in power, we can draw valuable lessons from Ui's struggle to empower those without a voice. We must question the role of the university in supporting the government—or increasingly, the market economy—in providing scientific "solutions."

Looking back on his battles against the political power that invited inaction and inversion in various forms, Ui highlights the structural failure of his research activities to secure promotion at the University of Tokyo. He remained "almost a faculty member" for twenty years, despite his significant contributions that bridged science and tacit knowledge while skillfully communicating episteme, *phronesis,* and perhaps a broader sense of *techne* to the public.

If the privileges and promotional opportunities associated with the university are removed, what remains? Regarding pollution, we realize that not much has been done to limit the activities of polluters, such as implementing technical measures to reduce pollution or developing alternatives. The victims of pollution have fought and sometimes lost their battles; in other cases, they have won, leading to a collective understanding of the progress made in the century-long history of pollution. This shared experience, Ui believes, can provide a new foundation for scholarly discipline in our universities (Ui 1971, 9).

Historically, systematized formal knowledge has been privileged, gradually diminishing alternative ways of knowing. We have adopted concepts such as efficiency and ecological modernization, which at first glance seem unrelated to politics. Both take "governance" as their starting point; however, if the status quo results from the accumulation of vested interests, then it is only natural that academia will become a "discipline for the powerful" (Ui 2014, 318). In his unfinished theory of pollution, Ui was aiming for a direction of study that serves the vulnerable.

To restore tacit knowledge, academia must transform into a problem-oriented endeavor. This transformation requires the knowledge and vocabulary to critique arguments that attempt to neutralize tacit knowledge. The belief in the superiority of formal knowledge has obscured the value of tacit knowledge. Yet tacit knowledge is not inferior knowledge. A good doctor can view their patient's symptoms both as data fitting a model and as the unique characteristics of each patient. While the doctor as a researcher is valued for their accumulation of formal knowledge about human beings in general, this should not lead to the neglect of the tacit knowledge they possess as a healer.

Formal scientific knowledge has indeed shed light on many aspects of environmental issues and has enhanced our understanding. However, as discussed in chapter 1, science does not dictate what to measure or analyze initially, nor does it clarify how to weigh assumptions and results as a society. These processes require judgment, a form of wisdom. Even if science aids us in making judgments and technology provides the means to act on those judgments, the need to explore "different ways of being" remains imperative. We must further investigate the nature of judgments themselves and how experiences are interpreted both before and after they become data.

The environmental state does not exist independently; it is supported by specific intellectual endeavors encompassing government, industry, academia, and civil society. The environmental state is supported not merely by individual policies and organizations, but rather by knowledge stratification. Following Ui's example, social scientists need to recognize the value of knowledge that cannot be captured through formal means. Social science plays a crucial role in envisioning alternative approaches to top-down scenarios that can produce inversions.

Given that most individuals who manage the environmental state are educated at higher-learning institutions, universities have a responsibility to protect, rather than exclude, problem-oriented and experience-based research. A movement in support of directing knowledge toward everyday problems, rather than generating more publications for scientific accolades, originated in Japan during a period of economic growth but never fully materialized. It is this lost art that we must rediscover in the forthcoming era of the environmental state.

What is the broad lesson from Ui's struggle in the pollution case in Japan? I believe it provides a framework of wider significance for examining other cases where power inequality exists between those who intervene and those affected by these interventions.

Whether the issue is pollution control or resource management, power gaps are always central to the outcomes, benefiting certain stakeholders while marginalizing others. The language of compensation often obscures the existing power inequalities that created and perpetuated environmental problems to begin with. Although the geographical locations and sectors discussed in the following chapters may appear distinct, the question of power inequality and the dependencies it creates remains central to our analytical efforts.

3
Exclusion: A Comparative History of Environmental States in Japan and Thailand

In certain countries, the seeds of inversion may have been sown much earlier, during the process that transformed the country into a modern nation and formed the relationships between nature and society. Intense interventions to nature can act as catalysts, prompting a developmental state to adopt the key characteristics of an environmental state. This chapter analyses the historical evolution of government intervention in Japan and Thailand, the only two countries in Asia that escaped colonization. We explore how the integration of peripheral natural resources into a centralized economy shaped and solidified the state-society relations in Japan and in Thailand.

3.1 The Dawn of the Environmental State

When forests are depleted, it is the people who bear the brunt of the consequences. Nowhere is this clearer than in Japan, where a constant stream of news about disastrous events—frequent landslides, wildlife control problems, and pollen allergies—can be traced back to mismanagement of the nation's forests. Once such problems arise, they are difficult to rectify, as damaged forests take generations to heal. We must coexist with the woodlands that remain while also striving to create new forests for the future. However, forests are not the only challenge. As discussed in the previous chapter, when a region attracts infrastructure to access its resources, that development initiates a cycle of investment and population growth, affecting land use planning in general.

Past development has repercussions for the future. But when does an investigation into the past become a futile endeavor? The further we retreat in time, the harder it becomes to draw a clear line connecting past events to the present. Take historian Conrad Totman's monumental work on Japanese forests, *The Green Archipelago: Forestry in Preindustrial Japan* (Totman 1998). Totman observed how these once-depleted forests, revitalized through careful planning and meticulous management by the shogunate authorities, showed remarkable recovery by the middle of the Edo period (from 1603 to 1868). However, the world during the Edo period was not nearly as globalized as it is today. Therefore, any lessons from that distant time may well be irrelevant now, especially for countries that are undergoing rapid economic development.

The second half of the nineteenth century is a more appropriate starting point for investigating the history of resource management in Asia, as it was around that time when nation-states began intense engagement with natural resources, forming a general "attitude" of resource governance.[1] A historical account allows us to uncover the roots of inversions and understand the apparently similar developmental stages where rapid political centralization took place. The late nineteenth century was also the period when many East Asian countries embarked on nation-building and state-led modernization campaigns. It marked the beginning of globalization, as systems and technologies originating in Western Europe were introduced in numerous countries. Nations scaled up their resource development and production systems to unprecedented levels, establishing policies to carry them into the modern age.

The focus of this analysis is to determine the extent of local residents' involvement—particularly their degree of inclusion or exclusion—in the process of state development of natural resources. This inquiry is essential because the inversions discussed here are closely linked to various forms of exclusion, such as the forceful removal of people from resources near their homes and the imposition of burdensome obligations on them. We cannot fully understand these mechanisms and the resulting inversions without being aware of how and why patterns of exclusion originated in the first place.

3.2 What We Can Learn from a Comparison of Japan and Siam

In the mid-nineteenth century, Japan and Siam (renamed Thailand in 1939) launched state-led programs for rapid modernization almost simultaneously. Both countries aimed to advance as quickly as possible to counter the ever-present threat of colonization by Western powers. Japan and Siam were in remarkably similar political situations. First, their modernization efforts extended far beyond the mere adoption of modern technology from Western Europe: Both countries sought to curtail the power of regional rulers by establishing centralized states.[2] Second, both succeeded in modernizing their systems of governance, which saved them from direct colonization. Both nations considered the mining and forestry industries to be keys to modernization, meaning that these two sectors played a central role in the development of state-society relations. Third, both countries opened their borders to world trade to promote rapid modernization and were willing to hire expensive foreign advisors to operate at the heart of their governments.

Even in the earliest days, both Japan and Siam were committed to the exploitation of their natural resources. This commitment required a restructuring of their relatively similar decentralized political systems and rigid class-based structures. To bring natural resources under the control of a centralized state, both countries swiftly established government departments to oversee resource management. However, their motivations for this shift differed. Why, during the early stages of their modernization efforts, did Japan and Siam place such emphasis on state control of natural resources? This chapter explores that question.

The early focus on natural resources is noteworthy, especially given that in the mid-nineteenth century, mines and forests contributed only marginal tax revenue. The economic history of Southeast Asia indicates that the colonial governance model emulated by Japan and Siam was primarily driven by the need for revenue. At the start of the Meiji period around 1868, land and alcohol taxes accounted for nearly 80 percent of Japan's total tax revenue. In contrast, in Siam, commercial services and royalties from tin mining and forestry generated only about 5 percent of tax revenue (Ingram 1971). During the reign of King Rama II (1809–1824), tin was the major mineral product but contributed less than 0.2 percent of Siam's annual revenue. During the Rama IV era (1851–1868), the combined income from tin, iron, and teak made up barely 0.7 percent of annual revenue (Wales 1934). While the accuracy of these figures can be debated, it can confidently be stated that nonagricultural resources contributed only a small portion of government income. This suggests that the government's objectives in seeking centralized control over these resources were not purely financial.

It is important to note that conditions in Japan and Siam differed in many ways, complicating straightforward comparisons. For instance, Japan was an isolated island nation with a sheltered history, while Siam shared land borders with neighboring countries that had already succumbed to colonial rule. Additionally, population density in Siam was much lower than that in Japan, resulting in less acute pressure on land and resource use. Moreover, at the inception of their modernization efforts, Japan had higher living standards and a more developed social infrastructure, and most Japanese citizens were literate. The requirement set by the Edo-era shogunate for regional rulers to maintain households in the capital, Tokyo, and to travel there created a well-developed communication system and an extensive road network. In contrast, the Siamese royal family's knowledge of the country's frontier towns was limited, and their contact with peripheral populations was infrequent (Bunnag 1977). In summary, Japan possessed significant advantages over Siam in its pursuit of centralization and expanded governance of its periphery.

The starting points for Japan and Siam were unique, yet both aimed for similar goals. Both countries sought to shift from decentralized governance, where regional rulers controlled the periphery, to a political system in which the central government had authority over all national territory. This process proved chaotic for both Japan and Siam as they attempted to extend central governance to the rich natural resources and diverse populations in the remote regions.

After achieving their nation-building objectives, however, both countries diverged sharply in their paths. Japan experienced remarkable economic growth but soon had to contend with the detrimental effects of large-scale industrial pollution. This prompted the creation of an environmental state that promoted relatively inclusive policies, allowing considerable leeway for both local people and private businesses. In contrast, Siam—renamed Thailand—took a more forceful and exclusionary approach toward its local populations. Making a closer examination of how these countries transitioned from one political system to another in the nineteenth century reveals the reasons for their differing approaches. In doing so, I aim to identify critical junctures or historical triggers of inversion in Japan and Thailand.

3.3 Centralization in Japan: The Story of Unavoidable Inclusion

3.3.1 The Impact of Land Tax Reform in Meiji Japan

Two key factors about Japan must be considered in our analysis. First, Japan's high population-to-land ratio has always been a decisive factor in its governance. Second, its regional rulers had a vested interest in managing the natural resources within their domains. Long before the Meiji reformers centralized political power in the mid-nineteenth century, feudal clans regulated the use of forests, water, minerals, and other local resources. Tax revenue and labor primarily came from farmers, who were legally obligated to cultivate specific plots of land and faced restrictions on their mobility. The Meiji Restoration marked the end of this rigid feudal system, profoundly transforming interactions related to resource control.

For our purposes, the 1873 land tax reform was one of the most significant and fundamental changes imposed by the Meiji government. It represented a sweeping attempt to transform governance from a highly decentralized, provincial rule-based structure into a system based on contracts between the state and individual landowners. However, the land tax reform extended beyond this transformation. It replaced Japan's old in-kind tax, primarily paid in rice, with a 3 percent monetary tax on anticipated profits from the land. Landowners received new land certificates that demarcated each plot's location, ownership, and estimated monetary value. This shift made it possible to sell land, which had previously been prohibited. The land tax reform was groundbreaking in that it enabled the Japanese government to create a fairer taxation system. Tax burdens that had once fallen primarily on farmers could now be directed toward urban residents and owners of forest plots.

Although farmland was the principal focus of the new land tax reform, the Meiji government also expropriated significant mines throughout the country, viewing these resources as essential to

Japan's modernization. Regional mines and industrial facilities vital for producing modern weapons and ships, for instance, were confiscated and placed under the authority of the Ministry of Public Works. Gold, silver, and copper were particularly valued, as they formed the foundation of the new government's finances and were crucial for establishing a nationwide currency system (Ministry of Trade and Industry 1966). Copper and coal could be exported for foreign currency, making them important during a time when Japan still relied on imports for all its modern machinery and equipment.

One overlooked aspect of the land tax reform was its impact on forests. As mentioned earlier, the contribution of woodlands to tax revenue was only marginal. However, forests covered about 70 percent of Japan's territory, and most arable land was adjacent to these woodlands. Thus, it was evident that the Meiji government's incursions into forest management would impact farmers in two important ways.

First, forest undergrowth was a primary source of fertilizer in the form of wood ash, as well as leaves and wood chips that could be composted. For farmers to adequately fertilize their crops from this source, they needed woodland areas roughly ten times the size of their fields (Takayanagi 2017). Consequently, they could not afford to have their access restricted. Any change in forest management policy could easily lead to existential crises for entire farming communities. Second, the reform aimed to differentiate between state-owned assets and privately owned land. Wooded areas produce various materials and are subject to preexisting local customs; however, the top-down reform left many sizable, forested areas in legal limbo. This uncertainty led to the ongoing conflict between the state and farmers who sought access to "ambiguous lands," much of which traditionally fell under the common lands (*iriai*) managed by the local community. Unlike agricultural fields, which had relatively well-defined boundaries that made ownership easier to determine, establishing a straightforward system of individual property ownership for forests proved to be much more challenging.

As mentioned earlier, most forested areas adjacent to fields had been managed communally by villagers as common property (*iriairin*). Many wooded areas were regarded as shared resources, granting access to farmers. Villagers could collect materials needed for agriculture and daily life at communally agreed times, gathering specified amounts using approved tools. The *iriairin* system ensured that farmers had the necessary resources (Ōta 2012) while also protecting the forests from overexploitation. These areas were difficult to incorporate into the new division between state-owned and privately owned land.

Few private woodland owners objected to the government's nationalization plans. Most forests had little commercial potential and were not viewed as favorable target for long-term investments. Japan's borders had been closed to foreign trade for centuries, resulting in an underdeveloped market for forest products. Most citizens were unaware that individual ownership of forests was even possible; consequently, the state's forest policy faced little opposition, particularly in comparison to its other land control efforts (Fukushima 1970).

However, the state's policy of centralizing the management of all woodlands with undetermined ownership quickly shifted. When the Meiji government realized it lacked the resources to effectively manage the country's vast new tracts of state forests, it decided to transfer and sell these lands to the private sector. The state had two primary goals in auctioning off what it considered unproductive wastelands: to generate revenue quickly and to expand its future tax base by encouraging new landowners to make the land productive.

This approach, consequently, deepened economic inequality in rural areas. It also intensified conflicts over resources and encouraged overexploitation, ravaging the forests. In 1879, when the state seized communal lands, approximately ninety-nine thousand Japanese farmers lost access to basic resources. By 1888, the number of affected farmers had risen to 1.8 million. Some of these individuals subsequently resorted to arson and illegal tree-cutting (Resource

Council 1960). The primary beneficiaries of the sale of woodlands were wealthy landowners. Increasing farmer riots and uprisings ultimately compelled the government to abandon its privatization policy.

3.3.2 Rural Resistance and the Forestry Bureau

In response to the escalating unrest and rapid depletion of forest resources, the Japanese state initiated efforts to reassert central control. In 1879, it established the Forestry Bureau within the Home Ministry.[3] Together with the Ministry of Economic Affairs, the Home Ministry implemented measures for domestic governance and the promotion of industrialization, a key policy goal of the Meiji government.

A pivotal figure in the establishment of the Forestry Bureau was statesman Toshimichi Ōkubo, one of the main champions of the Meiji Restoration. Ōkubo prioritized domestic stabilization over the pursuit of international interests and positioned forestry as a central pillar of industrialization. In his proposal for the Forestry Bureau's establishment, Ōkubo emphasized that "the protection of forests is a central task of the administration of a national economy" (Nishio 2021, 38–39), as forests served multiple environmental functions, such as protecting watersheds. He also recognized woodlands as essential to the success of state-coordinated projects in sectors like roads and shipbuilding. In short, the protection and maintenance of forests had to go hand in hand with other development efforts.

The government's centralization policy had implications that extended far beyond forest management. State forests became the primary source of timber for the nation, and anyone wishing to construct large-scale facilities—such as schools, hospitals, bridges, railroad beds, or ships—was required to obtain government permission to procure materials (Nishio 2021). The availability of hardwood from Japan's forests soon declined rapidly as factories and construction companies extracted enormous quantities of wood for use as fuel and building materials (Ōta 2012). The introduction of German forestry technology rendered indigenous forest management knowledge increasingly irrelevant. Across the country, a modern, centrally controlled system of ownership replaced the more traditional communal forest management practices (Niwa 1989).

Most forests were now under state control, but agricultural practices could not be changed overnight. People who had relied on forests for fertilizer mounted significant resistance throughout the country, demanding continued access to "their" woodlands (Fukushima 1970). While violent confrontations with the authorities were comparatively rare, resistance manifested through indiscriminate logging and arson, further accelerating the decline of the forests. Villagers developed sophisticated tactics to repel the state. Illegal loggers operated under the cover of darkness to fell trees, and when caught, they were backed by the promise that their families would be supported by fellow villagers. These individuals were willing to risk everything. Fire became another weapon for disgruntled locals; it was customary for the government to compensate villagers for burned trees and shrubs. This practice encouraged people to set fire to state-owned forests, watching as wildfires ravaged the land and refusing to cooperate with firefighters. In his pivotal study on farmer hostility, civil law scholar Mitsue Kobayashi noted that, in addition to their potency, these resistance efforts had an important side effect: They heightened villagers' awareness of their rights vis-à-vis the state (Kobayashi 1968).

Civilians who mounted this relatively behind-the-scenes opposition operated from marginal positions. While their actions hindered the government, they posed no actual threat to the system. Nonetheless, this relentless and protracted hostility made it increasingly difficult for the state to procure much-needed resources, including taxes and troops. Although day-to-day resistance does not deliver a knockout blow, it keeps an opponent under constant pressure, gradually narrowing their options. As farmers across Japan cleverly felled trees and set fire to woodlands, they exposed the limitations of the top-down control that the government was so eager to impose.

The new Forestry Bureau faced not only a barrage of attacks from angry farmers but also severe difficulties in securing the labor needed to manage state-owned forests. Ultimately, the department had to abandon its initial woodland strategy of exclusive state management. Instead, it began to develop more inclusive policies that accommodated local needs to some degree and incentivized villagers to provide much-needed labor (Ogino 1990). For example, the government introduced the "local facilities" (*jimoto shisetsu*) program to pacify restless farmers.[4] Through this program, responsibility for managing state-owned forests was transferred to local communities. In exchange for their labor, villagers gained access to resources such as wood for fuel and undergrowth for fertilizer. Consequently, most local facilities evolved into "commissioned forests" entrusted to local residents. This policy rectified the failures of earlier schemes that required contracts from forest users for the purchase of nontimber forest products (Matsunami 1920, 278). By 1926, 310 sites covering more than 40,000 hectares of state forest had been commissioned to local people (Kikuma 1980, 479). By 1937, in Akita Prefecture, for example, commissioned forests accounted for 70 percent of the state forests (Kikuma 1980, 492). The earlier suppressive policy had been replaced with a modern forest management concept influenced by German forestry technology, emphasizing the sustainable use of woodlands based on scientific evidence (Ogino 1990).

It should be noted that the Meiji government did not create the "local facilities" system solely to improve farmers' livelihoods. Its main purpose was to shift the burden of care for national forests to locals, eliminate theft, and motivate villagers to monitor wildfires. However, the new policy created space for state-society dialogue, leading the Japanese government to adopt more inclusive policies.

Another sign of inclusiveness emerged around 1908. Following the amendment of the Forest Law that year, the state began promoting "forest unions" based on guidelines from the Ministry of Agriculture and Commerce. These unions formed throughout the country to encourage the proper management of private forests (Matsunami 1920). Supported by low-interest government loans, the unions were responsible for practices such as thinning and branch trimming, afforestation, and public works aimed at soil conservation.

In short, resource management in Japan gradually shifted away from top-down policies that relied on police oversight to a strategy that encouraged locals to engage voluntarily in resource development (Resource Council 1960). Regardless of the motives, the state achieved its desired result: The country's forest acreage increased after the woodlands were returned to the care of villagers (Saito 2014).

3.3.3 State Intervention in the Mining Industry

Regulation of the mining industry followed a similar series of developments when the Japanese state sought to centralize mineral development. Japan has a longstanding mining tradition, with gold, silver, and copper being extracted in the archipelago since the ninth century. Before the Meiji Restoration, feudal lords had full control over mining activities in their domains. However, the Meiji government quickly brought all mines under state control by enacting the Japan Mining Law of 1873. This law stipulated that landowner rights did not extend to underground resources, ensuring that no ore could be extracted without government permission (Morita 2012). The centralization of the country's mining activities was deemed essential for securing the metals needed to mint coins, facilitating Japan's transition from a rice-based economy to a unified cash economy. Furthermore, the Meiji government was motivated by the quest for foreign currency, which could be obtained through coal and copper exports.

The administrative system governing mines underwent a series of complex organizational changes, culminating in the establishment of the Department of Mines in March 1886. In the early days of the Meiji Restoration, the government attempted to impose unified control over copper production, a major export for the central city of

Osaka. Due to the substantial initial investments required, jurisdiction over mining was initially granted to the Ministry of Finance. Over the subsequent years, control shifted to the Ministry of Agriculture and Commerce and then to the Ministry of Civil Affairs and the Ministry of Public Works before ultimately being restored to the Ministry of Agriculture and Commerce in 1886.

As this shifting indecisiveness indicates, the central government struggled to effectively administer the newly acquired mines, and bureaucrats ultimately failed to achieve the anticipated increases in mineral production. Inefficient production processes and high costs associated with employing foreign technicians compelled the state to privatize key mines, beginning in 1889. But problems continued to plague the mines. An economic downturn and increasingly harsh working conditions sparked a series of labor disputes across Japan, culminating in social unrest that reached a boiling point in 1907. Miners rioted at major copper mines in Ashio, Beshi, and other locations, bringing operations to a halt. In response, the government deployed police to suppress the strikes (Ministry of Trade and Industry 1966, 396).

Between 1869 and 1898, there were seventy-one labor strikes in Japan, with twenty-eight (40 percent) occurring in mines, highlighting the harsh and dangerous working conditions faced by miners (Sasaki 1971, 252). In the early modern period, most miners were farmers who only worked underground during fallow seasons. As working conditions grew increasingly grim, the government found it more difficult to attract enough laborers. Moreover, most workers willing to endure the mines were part of a "spillover population," outsiders to the four major social classes of Edo-era society: samurai, merchants, farmers, and craftsmen. This affected all labor management policies the government could realistically implement (Yoshiki 1980, 4). Ultimately, significant mines, including the Miike coal mine, addressed their labor shortages by employing prisoners and Korean forced laborers, both of which were readily available even during the peak agricultural season. Although mines were a central element in the government's industrialization efforts, securing a steady supply of labor proved challenging.

As previously mentioned, Japan has had centuries of mining tradition. This posed complications for the Meiji government, as all easily accessible mineral deposits had long been exhausted. Major extraction operations now had to be conducted deeper underground and farther away from population centers, which generated new demands for skills. Since the Edo period, prisoners had performed simple manual tasks in the pits, such as water drainage and transportation. Advanced tasks, including refining and sorting, required technical skills that the prisoners did not possess. When Japan adopted Western technology and mechanized many processes, labor-intensive jobs like water drainage were also automated. Consequently, prisoners were relegated to digging and transporting materials in the deep pits—dirty and dangerous jobs that were unpopular as well as hard to automate. Amateur artist Sakubei Yamamoto worked in the coal mines around Chikuhou in Fukuoka Prefecture from the end of the Meiji period until after World War II. He left a unique record of conditions in the mines through vivid ink and watercolor illustrations (see figure 3.1).

The first exposé of the extreme and inhumane working conditions in the mines was made public in 1889. That year, Meiji- and Taisho-era journalist Koichi Matsuoka conducted a "field inspection" of the Mitsubishi Takashima coal mine in Hizen. His subsequent report caused a public uproar, prompting the government to launch a full-scale investigation. This led to the Mining Regulation of 1890, which was based on the Prussian Mining Law and became the first official acknowledgment that the health and basic rights of miners should be protected. Notably, the Mining Regulation was promulgated no less than twenty years before the Factory Act of 1911, which recognized the rights of factory workers.

It would be naive to assume that this regulation signaled a genuine increase in government goodwill toward laborers. The

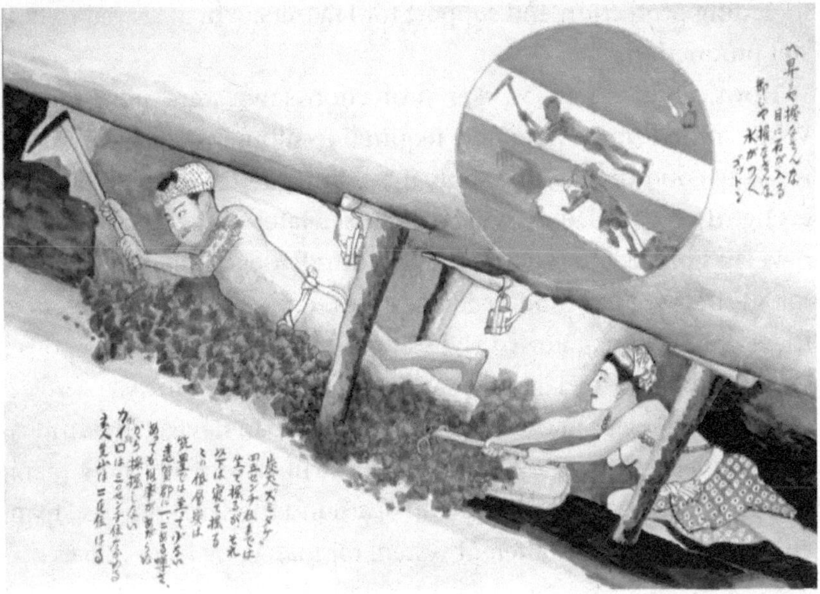

Figure 3.1
Illustration of working conditions in mines, which remained largely unchanged until the early Showa period (Sakubei Yamamoto, Nebori no. 38). *Source*: @Yamamoto Family, Tagawa City Coal Mine Historical Museum.

government's primary motivation for improving harsh working conditions was to decrease labor turnover. Meiji bureaucrats were aware of the negative aspects of Western industrialization, recognized the damage that labor disputes could cause, and sought to prevent these conflicts from arising in the first place (Garon 1987).

In summary, following the Meiji Restoration, Japan underwent rapid policy shifts concerning centralization and privatization. After years of turmoil and resistance from farmers and workers, the government was forced to acknowledge the limitations of its initial centralization approach, which relied heavily on police enforcement. The inhabitants of Japan's resource-rich peripheral regions managed to maintain leverage in negotiations with the state, and both the Mining Law and the Forestry Law could not ignore these groups. On the contrary, these laws reinforced a range of rights, including protection and support for laborers, which served to bolster production.

However, the new worker protection laws were no panacea. While mine owners were now required to submit regular reports on working conditions to the state, they continued to keep their workers tightly controlled. As will be discussed later in this chapter, these new laws made laborers and farmers utterly dependent on the state for all their security and health services. From another perspective, these policy shifts allowed the state to extend its regulatory power to encompass all social classes and regions of the country.

This concludes our examination of how Japan navigated its industrial modernization process. Next, we will investigate how Siam, which began its modernization drive around the same time as Japan, ended up with a very different system for managing its resources.

3.4 Centralization in Siam: A Story of Exclusion

3.4.1 Tin in the South and Teak in the North

Until the second half of the nineteenth century, Siam had a decentralized system of governance similar to that of Japan during the Edo period. Geographical regions in Siam were managed by feudal lords known as Chao Muang, who wielded considerable influence over their territories. Local citizens paid taxes and provided corvée labor to these lords in exchange for protection and recognition of their usufruct rights (Sethakul 1989). The feudal lords maintained a particularly tight grip on resource management. Teak, which had significant commercial value, was abundant in the north, while the south boasted rich deposits of tin, a key raw material for early factories and exports. The government's centralized control over these resources marked a significant political shift, as Bangkok asserted its authority over the local lords by transforming resource management.

International politics significantly influenced Siam's efforts to reform its domestic political system. For instance, in 1855, the British signed the Bowring Treaty with the Siamese government to promote free trade. This unequal treaty granted the British the ability to trade freely at fixed low tariffs, full extraterritorial powers (such as exemption from prosecution in Siamese courts), and the right to own land in Siam. Aggressive moves by Great Britain on its western borders and France to the east compelled the Siamese government to take action. To effectively resist the threat posed by Western colonial powers, Siam needed to establish a centralized bureaucracy supported by a unified tax system and to overhaul the feudal lords' control over local resources.

Unlike population-dense Japan, Siam was sparsely populated and faced a general shortage of workers, which made control of the labor force rather than the land the government's top priority. Siam had a well-developed system of property rights concerning people (e.g., slavery) that enabled a per capita tax. Until the end of the nineteenth century, debt enslavement remained common, and ethnic minorities frequently suffered from slave raids. By the mid-nineteenth century, slaves constituted about a fourth to a third of Siam's total labor population (Thompson 1947). However, in terms of managing land ownership, the country lagged behind (Feeny 1989). Some observers theorized that the scarcity of labor hindered the government's ability to develop Siamese resources to their full potential (Wales 1934).

One of the key administrative reforms initiated by King Rama V, who ruled from 1868 to 1910, was the transition of Siam's labor-based governance system to a tax-based one. This shift enabled the monarch to diminish the power of feudal lords and establish a central government (Feeny 1989). The government also introduced a monetary tax scheme, similar to Japan's approach after dismantling its in-kind tax system. The registration of land tenure, which formed the basis of this new tax system, began following the Land Title Deed

Issuance Act of 1908. At that time, the state showed little interest in land beyond rice fields. Consequently, most newly recognized tenure rights were concentrated in the central plains, where the majority of the country's rent-earning rice fields were located (Ingram 1971).

Siam's eventual interest in managing its resources was largely a response to the encroaching actions of the British. The central government in Bangkok felt powerless against the feudal lords, who granted British companies rights to log teak and mine tin. European companies started making inroads into the region in the 1860s (Ross 2014). While tin deposits were scarce in Europe, they were abundant in Southeast Asia, particularly on the Malay Peninsula, and by the turn of the twentieth century, demand for tin was high. It was used in weaving, electric machinery and weapons production, and food canning and as a corrosion-preventive for iron. In nineteenth-century Siam, the royal family had an official monopoly on tin exports; however, in practice, much of the output was controlled by a small number of Chinese families connected to influential provincial governors in the southern regions (Falcus 1989).

The tin industry, which was primarily run by Chinese and British companies, experienced relatively few territorial conflicts. In contrast, the teak trade in the north presented significant challenges. Reginald Heber Macaulay observed that the Siamese government's main concern was not so much with logging itself but with governance issues that impacted its domestic and international dealings: "The growing importance of the timber trade was no doubt one of the reasons which determined the Siamese government to take more control of affairs in the north. Their idea no doubt was to get control of the Laos States of Northern Siam, but they proceeded gradually, not wishing at first to upset the reigning chiefs, though eventually they expropriated them and took the administration of Upper Siam entirely into their own hands" (Macaulay 1934, 59).

Woodlands long played a significant role in the nationhood of ancient Siam. As survey-based maps of premodern Siam did not exist, woodlands served as effective, though imprecise, borders between sovereign states (Thongchai 1994). The only clearly delineated borders were around timber and tin-mining sites of interest to the British (Thongchai 1994).[5] Efforts to control the country's resources were driven not solely by the market value of teak and tin but also by their geopolitical significance.

In 1891, the state established the Royal Mines and Geology Department (later renamed the Mining Department) as a means to constrain local actors and impose control from Bangkok. The 1901 Mining Act, drafted by German and British advisers, established a regulatory framework for the mining industry. One of the Mining Department's first directors was the British geologist Herbert Warrington Smyth. In his memoir, *Five Years in Siam*, Smyth lamented the lack of a formal apparatus to administer the country's mines: "There were a dozen or so big mining concessions in existence, covering in some cases a hundred square miles, a weariness of the spirit of their owners, on which, for the most part, no rents had been paid and no work had been done. They had been mostly granted to men of the concession-hunting type, whose sole objective was to sell their concessions as soon as possible for the highest price to some gullible company" (Smyth 1898, 33).

King Rama V introduced a licensing system in an effort to resolve the ongoing disputes between mining companies, provincial governors, and Bangkok (Krom Sapayakorn Tarani 1992). The new Mining Department immediately began conducting geological surveys and drafting a code of mining regulations to ensure a steady flow of tax revenue. This technology was also crucial to the establishment of a centralized Siamese state (Loos 1994).

The primary laborers in Siam's mines were Chinese immigrants (Ingram 1971, 210), who possessed specialized mining knowledge. Working conditions were harsh, prompting companies to offer high wages to retain workers (Skinner 1957, 110). Although the Chinese were secretive about their technical know-how to maintain their advantageous positions (Malloch 1852, 22), the government made no systematic efforts to train local mining specialists or replace Chinese laborers until after the 1930s, when a powerful nationalist sentiment emerged in the country (Krom Sapayakorn Tarani 1992, 183–92). Small Chinese mining companies paid royalties of 10–16 percent to the government (Ingram 1971). During the modernization period, the government exerted only limited indirect control through its leasing of mining rights, while tin mining in the south remained largely an economic enclave dominated by the Chinese.

After the British exhausted teak forests in neighboring Burma, they moved into northern Siam, where they gained significant control over the exploitation of the region's abundant forests. Alarmed by the ravaged teak forests of Burma and the expanding British influence in the region, the Siamese government quickly centralized the management of the country's forests (Barton and Bennett 2010). The Royal Forest Department was established in 1896 under the Ministry of Interior. Its first chief, known as the Conservator (*Chao Krom Pamai*), was the English forester H. Slade, who had previously worked in forest management in Burma. He held this position until 1901. In its early years, the primary responsibilities of the Siam Royal Forest Department included survey work and inspections in collaboration with the Royal Survey Department. This involved demarcating lease boundaries, identifying tree species for cutting, and settling disputes (Slade 1901). To limit excessive logging and allow Siam's forests a chance to recover, Slade suspended the issuance of logging leases for at least six years, marking the first systematic efforts in Thailand to manage forests (Slade 1897).

By around 1900, the Siamese government began to recognize the political significance of its timber and minerals. The establishment of both the Mining Department and the Royal Forest Department within the influential Ministry of the Interior underscored their importance.

3.4.2 The Seeds of an Environmental State in Siam

Two developments in Siam bolstered the emergence of an environmental state. First, the government leveraged the threat of Western interference to legitimize its efforts to centralize resource management. Macaulay noted, "After a short time a regular Forest Department was established. The Siamese Government justified the step to the chiefs by implying that with this measure the Europeans would be kept in line" (Macaulay 1934, 60). Second, by articulating an agenda for resource governance, the government created a blueprint for a new relationship between the state and society.

The Forest Department soon declared that any land in the north without legal ownership or usage was considered the property of the king and the Forest Department (Mekvichai 1988). Policymakers were influenced by the prejudice that locals, particularly hill people who practiced slash-and-burn agriculture, were a hindrance to the sustainable management of Siam's forests. Slade, the first head of the Royal Forest Department, characterized the agricultural practices of hill people as "wasteful." According to Slade, if the government failed to regulate the hill-clearing tactics of locals, it would never ascertain the amounts of teak at its disposal (Slade 1901, 7). In this way, modernizing states claimed a monopoly over resource management while disregarding the customary usage rights of forests (Peluso and Vandergeest 2001). Consequently, tensions over forest use resulted in a permanently antagonistic relationship between the people and the state.

When a state takes steps to protect forests within its borders, conserving nature is typically not its primary motive. Based on research into primary historical sources, Atsushi Kitahara noted that in the 1920s, when foreign capitalists seized large tracts of Siamese land, the government introduced the concept of "public land" to protect and defend its territory (Kitahara 2012). Around the same time, when a serious economic recession hit Thailand, the country's resources faced the risk of overexploitation, and the only solution was state enclosure of land (Kitahara 2012, 329). These nationalization efforts had a significant impact on the relationship between the state and the communities living near the sought-after resources. Tin and teak, the resources most coveted by Bangkok, were plentiful along the nation's northern and southern borders—areas that overlapped with the traditional lands of minority ethnic groups, who were considered obstacles to national unification.

In conjunction with its efforts to secure the country's timber resources, the Forest Department also established distribution channels and communication infrastructure to engage with rural people. The department's work was invaluable to the state as it extended its power to the periphery. This centralized management of the country's forests was also an attempt by Bangkok to solidify its authority over rural chiefs. Forest administration became an important tool for diminishing the economic base of the once-powerful feudal lords. In the words of Macaulay,

> For generations these chiefs (in Chiengmai, Laon, and Nan) had been allowed to govern the country within their respective states without any control from Bangkok, though in theory their power was circumscribed. Therefore, it was not an easy matter to upset the old regimes suddenly, with the reigning chiefs still alive. It also happened that all the chiefs were old men at the time the Siamese Government started its scheme of assuming control of the Laos States. This was around 1896 and coincided with the advent of H. Slade to start a Forest Department. Thus, it may be truly said, that it was through the Forest Department that the Siamese Government inaugurated the system, and this ended with their securing absolute control of the Laos States. (Macaulay 1934, 65)

As state-owned forest reserves expanded, they sparked acrimonious conflicts with villagers who had traditionally relied on the woodlands for their livelihoods. This resistance left the government struggling, leading it to grant usage rights on an ad hoc basis (Kitahara 2010). Initially, the state implemented a clear-cut policy of "dos and don'ts," but over time, it softened its approach, allowing the collection of forest products for subsistence needs and designating areas that could be accessed without permission.

However, this easing of restrictions was merely a passive acknowledgment by the state that peripheral communities required access to forests to support their growing populations. It was not until the 2000s that the government developed affirmative policies recognizing local rights in state-owned forests.

3.5 How Siam and Japan Developed Distinct Patterns of State-Society Relations

For Siam, natural resources represented more than just commodities or sources of income; they were the primary means through which state power penetrated peripheral communities. From this perspective, it is understandable why the government made significant efforts to extend its control over remote resource sites that initially held little monetary value. As foreign capital continued to infiltrate Southeast Asia, King Rama V grew increasingly concerned about the scramble for concessions, and the rapid depletion of Siam's northern forests justified the government's attempts to impose a "science-based" management of resources in the periphery (Barton and Bennett 2010, 75).

Around the same time, Japan faced a different problem: Resources were becoming alarmingly scarce. In response, Toshimichi Ōkubo, the mastermind behind the Forestry Bureau, advocated for involving local people in protection activities. Ōkubo's recommendation stemmed from the recognition that the state alone could not effectively manage its forest resources. The government understood that it needed assistance to care for this precious resource, which motivated its support for the formation of forest unions around 1900.

While Siam and Japan pursued modernization through similar methods of nationalizing peripheral resources, there were two key differences in how these countries evolved into environmental states. First, the relationship between private business and the government differed significantly in each case. In Japan, the government quickly reversed its centralization policy and auctioned off key resources to the private sector (albeit only to state-sponsored businesses). In contrast, Siam did not sell resources to private companies. Instead, it tightened central government control in response to rampant logging and excessive allocation of mining rights. Although some Chinese companies became powerful mining stakeholders, no domestic actors managed to dominate the exploitation of the country's most important national resources. This sharply contrasts with Japan, where industrial and financial conglomerates, known as *zaibatsu*, thrived through resource development.

A second key difference is evident in how these two countries developed their labor forces. The Japanese government was proactive in offering welfare services to laborers and enacting legal instruments for their protection. For example, in the forestry sector, the government revised the Edo-period system of "shared forests," which involved contracting private actors to plant forests on government land and share the profits with the state. Villagers were granted free access to nationalized forests to gather undergrowth, providing labor in exchange for the management of these same forests. In the mining sector, the Japanese government established policies to protect miners' rights. In Thailand, by contrast, communal rights to forests were not even debated until the 1980s, and no legal framework existed to protect miners until after World War II. Throughout Thailand's modernization process, the central government consistently enforced a policy of control and allocation of resource access (Sato 2003).

It is notable that during Japan's modernization, labor organizations among miners were strengthened rather than marginalized. Middlemen and subcontracting labor bosses were key players in Japan's mines, and their presence meant that the benefits promised by the government's worker protection policies were not always realized.

The result was twofold: Labor movements sought direct support from the state to bypass industrialists and labor bosses, and workers established self-help organizations.

One noteworthy example is the *tomoko* mutual aid organizations, which were widely observed in many coal mines across Japan and functioned as informal networks for skilled miners (Matsushima 1978). *Tomoko* provided insurance for individual workers in cases of injury or illness. Mines were often abandoned once all accessible ore had been excavated, forcing employees to relocate. *Tomoko* assisted skilled miners in moving from one place of employment to another, thus helping to disseminate technical knowledge throughout the country. To improve their technical skills, some miners preferred to transfer between mines (Murakushi 1999). *Tomoko* was instrumental in supporting these relocation efforts and sought to ensure respectful treatment of skilled workers at their new jobs. The widespread existence of such systems, which helped skilled workers pool resources, attests to the shortage of expert labor in Japan and the harsh working conditions in the mines.

In Siam, Chinese migrant workers laboring in the country's mines also established self-help organizations. Unlike the Japanese *tomoko*, which were recognized by the government, Siamese mutual aid organizations were more akin to secret societies that operated in defiance of the authorities (Cushman 1991). While revolts occasionally disrupted the government's administrative reforms, such occurrences were rare, and the government often succeeded in co-opting them with financial incentives (Ramsay 1979). In the south, the dominant Chinese mining conglomerates wielded considerable influence, taking it upon themselves to settle disputes. For the government in Bangkok, these conglomerates served as a useful buffer against British influence on the Malay Peninsula (Cushman 1991). Protection for laborers was not codified into law until 1956, at which point workers gained the right to organize and bargain collectively with employers (Mabry 1979). In short, laborers in both countries worked to establish self-help organizations. On the surface, these organizations supported similar goals, but in Japan, they enjoyed a much more favorable relationship with the state than in Siam.

3.6 What Made Japan and Siam Different?

3.6.1 Socio-Ecological Endowments

Japan and Siam began their modernization processes around the same time, with the mining and forestry industries playing pivotal roles in shaping a new relationship between the state and society in both countries. However, natural resource development yielded very different outcomes in Japan and Siam for three main reasons.

First, in Siam, mining and forestry operations occurred in enclaves that were geographically distant from the rice-cultivating centers of the country. Resource development relied on the labor of hill people and Chinese workers, groups that functioned outside the direct control of the central government. During the 1920s and 1930s, when nationalist sentiment was strong, the state had little incentive to improve the working conditions of these non-Siamese laborers (Thompson 1947, 230). In contrast, Japan's forest products and minerals were essential for peasant agriculture, which generated most of the nation's tax revenue. Consequently, Japan's resource development fostered a closer relationship between the state and society. The government's decision to grant local communities a degree of control over forest management contributed to the preservation of forest coverage. Additionally, Japan already had a solid foundation from the Edo-period forest cultivation system, which relied on local foresters and private companies (Saito 2014). Japan also faced a severe labor shortage, prompting the state to improve working conditions and acknowledge customary worker organizations, such as the *tomoko*.

The second reason for the differing outcomes in Japan and Siam was their approach to incorporating Western technology. Siam employed Europeans in high-ranking positions within departments responsible for resource development. These Europeans brought with them the notion that resources should be exploited as efficiently as possible, often disregarding local communities and customs. In Japan, however, power was primarily held by domestic politicians and reform-minded bureaucrats who had studied abroad. While eager to learn, they also sought to adapt Western knowledge to their own context and were aware of the negative side effects of Western institutions and technologies. In the Home Ministry, "social bureaucrats" recognized the harmful consequences of industrialization in Europe and played a key role in enacting protective measures for factory workers (Garon 1987).

Third, past experiences with resource development took different forms in Japan and Siam. Japan had a longer history of resource exploitation, encompassing a wider range of resources, some of which were already scarce by the early Meiji period. This scarcity motivated people to consider how to make the most of what they had. These circumstances fostered a high level of interdependency among resources, exemplified by farmers who relied on forest undergrowth to fertilize their crops. In Japan, this interdependency became one of the foundations of state-society relations. In contrast, early forest management in Siam was primarily focused on securing commercially valuable timber, such as teak. Mining development in Siam concentrated almost entirely on tin, reflecting the Siamese government's limited interventions in the sector. In Japan, however, government involvement in mining operations targeted a much broader range of minerals, including copper, gold, silver, coal, and iron. This prepared the Japanese state to adopt a more comprehensive approach to environmental management.

A state manages to penetrate its remote societies to varying degrees, and this variability cannot be attributed solely to administrative capacity. The comparison of Siam and Japan illustrates that the state's capacity to reach communities beyond the center is influenced not only by government actions but also by the specific ways in which the state and societies interact. In Japan, state involvement in mining relied on privatization and traditional workers' associations. In Siam, as the country modernized, the state's initial incursions were exploitative, rooted in a colonial approach that treated natural resources as tradable commodities; consequently, the state's influence remained limited outside labor- and resource-rich areas. The interactions of Japan and Siam with their populations through resource management were distinct, ultimately leading to two different patterns of state-society relations. These contrasting systems guided the evolution of each country into a unique type of environmental state.

3.6.2 Sowing the Seeds of Inversion

Japan and Siam advanced two different models of state involvement in natural resource development: an inclusive approach and an exclusive one. However, as this chapter has demonstrated, these divergent strategies did not necessarily develop through deliberate choices made by their respective governments. In Japan, the government increased its control over state-owned resources by engaging local populations in governance, not out of a desire to prioritize the welfare of those involved but because limitations in the state's administrative capacity left it with no choice but to rely on locals. The government had to adopt an inclusive approach; the alternative would have meant abandoning its modernization ambitions.

The situation in Siam was complicated. Its exclusionary resource management policies effectively hampered the Siamese government, limiting its reach. In some ways, this allowed local people greater freedom in the early stages. However, the interdependency with the state gradually expanded as Thai modernization penetrated deep into the remote periphery.

One fact is certain: The governance of the environment in Japan and Siam has been closely intertwined with the governance of the people. When these states sought to concentrate power over peripheral resources in the political center, their efforts shaped state-society relations in these remote regions for generations. Both countries differentiated land as either private or state property and appropriated nearly all underutilized forests for state use, leaving a lasting legacy of conflict between the state and the people. At this point, the methods employed by Japan and Siam diverge. The Japanese state quickly adapted its early approaches to accommodate the people at its fringes and enlisted them in the state's policies—such as recognizing the customary use of forests and facilitating the creation of forest unions. In contrast, the government in Siam adopted a resource management approach that continues to disregard the hill people and local communities.

In this chapter, we investigated why early modern Siam and Japan aimed to exert control over their natural resources at a time when forests and minerals contributed little to state revenue. We discovered that the process of creating wealth by developing and utilizing resources shapes relationships among people and that these new relationships can alter the character of a state and its approach to governance. The historical analysis conducted in this chapter helps us appreciate the divergent paths of resource governance, contrasting them with the convergence that many early modernization theorists expected latecomer countries to achieve. It also demonstrates why history is important for understanding the evolution of inversion (Cullather 2000).

Political leaders in Japan and Siam faced similar threats and responded to the needs of modernization, leading to divergent outcomes regarding state-society relations. Both countries faced the looming threat of colonization, experienced power struggles between the political center and the periphery, and witnessed conflicts between farmers and the ruling class. Despite these similarities, Japan adopted an inclusive approach to resource management, while Siam opted for a more exclusive path. In Japan, the state and the periphery depended on each other to manage resources effectively; they had no other viable options. As we will see in Chapter 6, this limited latitude for choice is solely the perspective of policymakers, not that of the general public. Here, we may discover leverage for the future to develop environmental states that can minimize inversions.

4
Release: The Cambodian Government Relinquishes Control over Fisheries

Inversion is difficult to predict, especially when conservation policies are sugar-coated with slogans like "community empowerment" and "poverty alleviation," which imply a departure from state centralization. A pertinent example is a policy implemented in Tonle Sap, the largest freshwater lake in Southeast Asia, located in Cambodia. Long known for overexploitation and uncontrolled illegal fishing, the government implemented a radical fishery reform in 2012 to abolish the fishing lot system that had been in place for over a hundred years. This chapter analyzes the gap between the rhetoric and the reality of fishery conservation. It explains how delegating governance to communities can lead to inversion when responsibilities are unclear and the affected communities lack the necessary resources for effective management.

4.1 Enclosing a Nonstationary Resource

4.1.1 The Cradle of Life

Tonle Sap is the largest freshwater lake in Southeast Asia, providing Cambodia with a range of crucial economic and social resources.

Figure 4.1
Stilt houses in Kampong Phluk on Tonle Sap Lake. *Source*: Photo by the author (November 26, 2024).

An estimated 1.7 million people, spread across over 1,500 villages, make their homes on and around the lake (Yoshimura et al. 2022; Sithirith 2014, 597). A distinctive feature of Tonle Sap is the fluidity of its boundaries; during the rainy season, the lake swells to several times its dry season size. Lake dwellers relocate as the shorelines shift, dividing their time between farming and fishing with the changing seasons.

There are many reasons why lake dwellers choose to live close to the water. Community members have long benefited from the freedom to live and work on the lake, with many necessities available to anyone with a boat. The climate is favorable year-round, and fish are—or at least used to be—abundant, creating economic opportunities and reducing the risk of famine. Communities across the lake also contain schools, health centers, and other essential social infrastructure necessary to sustain their livelihoods.

However, the region's peaceful atmosphere can be deceptive. The abundance of resources, combined with severe poverty, makes Tonle Sap Lake an ideal habitat, but it can also lead to conflicts over access to areas with bountiful fish harvests. When Cambodia was under French rule beginning in 1863, the colonial government established a licensing system to regulate access to specific fishing sites. This system involved partitioning valuable fishing grounds and auctioning off the rights to operate fisheries.

This practice led to the creation of "fishing lots." Fishers who successfully bid for these lots sought to protect their investments by enclosing productive areas with fences. Nonetheless, illegal fishing that ignored the boundaries persisted. To complicate matters, fishers and farmers often competed for the same space; at times, farmers would cut down sections of the flooded forest around the lake to secure more arable land, thus reducing the fish habitat (Ratner et al. 2017, 73). Various reports on the mismanagement of fishery resources in Tonle Sap led the government to push for a fishery reform starting in October 2000 with the following emphases: (1) strengthening and expanding community fisheries (2) empowering and giving ownership to local communities to manage fisheries (3) conserving natural resources, and (4) promoting the development of aquaculture (Cambodia Abolishes Fishing Lots, 2012, 13).

On March 12, 2012, the Cambodian government stepped even further to announce the entire abolition of the century-old fishing lot system. This change has had profound effects on the economically and environmentally valuable ecosystem of Tonle Sap Lake. Although the government began reducing the total area of fishing lots since the reform began in 2000, it was not until 2012 that the system was completely abolished. While this shift was initially welcomed by small-scale fishers relying on Tonle Sap's fishery resources, it has led to a rapid decline in fishing stocks, as well as the conversion of seasonally flooded forests—critical habitats for fish—into rice fields (Mahood et al. 2020). This decline has been exacerbated by other problems linked to climate change, intensifying El Niño, which has resulted in reducing the fish size and diminished biodiversity in Tonle Sap (Seiff 2017).

This chapter examines how the dismantling of the fishing lot system—a seemingly sensible approach to local empowerment and conservation—has triggered inversion in Cambodia. It focuses on the relationship between the state and society through resource manipulation and explores how they influence the perceptions of the natural environment as a subject of exploitation. Studies of resource management in Southeast Asia typically investigate how governments enclose and monopolize resources under the guise of economic development, employing both conservation and privatization as key tools (Hall et al. 2011). Such state actions are often referred to as "green extractivism," a large-scale exploitation of natural resources under the guise of environmental sustainability (Bruna 2022), and tend to exacerbate conflicts between authorities and residents, often to the detriment of local interests, particularly those of Indigenous communities (Jamin and Hermawan 2024).

In contrast, the abolition of the fishing lot system at Tonle Sap Lake stands out as an unusual example of a government policy that was met with jubilation by the locals. On the surface, this approach seems to have reduced state dominance and reinstated certain public freedoms, allowing small fishers to have access to previously prohibited domains. However, this policy shift also signals a trigger for inversion. As I will argue, the case of Tonle Sap Lake fisheries exemplifies a different mechanism of state power, where regime legitimacy is gained at the sacrifice of the resource base.

4.1.2 The Territorialization of the Commons

Let us review the evolution of theories surrounding common resources, especially fisheries, since it gives us a clearer vision of the core issue of resource management. A key distinction between fish and other material resources is the difficulty of preserving them. Consequently, like water, fish from the lake are widely consumed locally, while those with access to refrigerating facilities would have the opportunity to export. In contrast, wood and minerals are logged or mined specifically for transport to other regions and, unlike fish, sustain minimal damage during transit. Recently, however, the advent of new freezing technologies for preserving fresh catches has enabled fish to be commodified. Fish are also notable as a critical source of protein for humans, especially in the Mekong region of Southeast Asia: Fish accounts for 81 percent of the population's protein intake in Cambodia and 48 percent in Laos (Baran and Borin 2012).

The management of fish stocks is often cited as a classic example of the challenges associated with the "tragedy of the commons." In the scenario formulated by ecologist Garrett Hardin, there is no technical solution to collective action problems, as the outcome stems from individual rationality (Hardin 1968). If the benefits of overfishing outweigh the costs of degrading the fishing grounds, it becomes rational for each fisher to continue fishing without restraint. In this context, addressing the commons problem may lead to the notion that "injustice is preferable to total ruin" (Hardin 1968, 1248), where mutually agreed-upon coercion is justified.

Long before Hardin published this seminal paper, scholars were already analyzing the issues inherent in fisheries as commons (Gordon 1954). By definition, the commons are shared resources that must remain accessible to all potential users. Existing natural resources, such as forests and fish stocks, are viewed as local commons not only because their continued existence as shared reserves is vital to a community's survival but also because excluding certain groups from accessing these resources would be too costly to implement.[1]

The state has primarily relied on two options to ensure the sustainability of such shared resources: market mechanisms and direct regulation through various policy instruments. The advantages and disadvantages of both approaches have been extensively studied and debated. The first approach, often referred to as "leaving it to the market," generally adopts a neoliberal perspective. In practice, some local resources are commercialized to encourage the communities dependent on these reserves to take an interest in their conservation (Castree 2008a, 2008b). For example, in Southeast Asia, scholars and policy practitioners have linked the commercialization of natural resources—such as cash crops like coffee, wood products, mushrooms, and bamboo shoots—to the connection between forest conservation and income generation (Nevins and Peluso 2008).

In the second approach, the state explicitly enforces its conservation objectives by enclosing the resources it aims to protect, typically seen in forests and land (Hall et al. 2011). This approach also includes relocating people to shield them from health risks. For example, in the Kampong Chhnang district of Tonle Sap, nearly five thousand families from floating communities were mass-evicted and relocated to land to avoid exposure to poisons and pesticides used in fishing and agriculture. The ingestion of such substances contributes to health problems and increases water pollution levels (Smith 2021, 294).

In reality, however, the two approaches are mutually interdependent. Large-scale commodification of shared natural resources necessitates some form of state involvement, particularly in property assurance. Top-down regulation, on the other hand, often requires economic compensation based on the market value of the houses and land and the assurance of income-generating opportunities to offset the burden imposed by the regulation. Researchers have observed that the most "successful" resource commercialization often develops only after resources are enclosed and local populations are excluded (Dove 1993). This exclusion, enforced through state-mandated enclosure, is known as "territorialization" (Vandergeest and Peluso 1995). Territorialization is "about excluding or including people within particular geographic boundaries and about controlling what people do and their access to natural resources within those boundaries" (Vandergeest and Peluso 1995, 388).

In essence, territorialization describes a process through which the state effectively colonizes land within its borders, including land and forests (Peluso and Vandergeest 2020).[2] In some countries, like in Cambodia, such state-confiscated lands are then redeveloped to be sold to large-scale companies for business purposes. In Phnom Penh as of 2020, for example, 2,004 developers acting with government backing destroyed the 1,500 hectares of wetlands, filling them with sand dredged from the Mekong and Bassac rivers to prepare for the construction of a vast 2,500-hectare satellite city, dubbed "ING City," the largest development in Cambodia (Knaus 2020). This is a typical way that state and market interact to territorialize nature in preparation for inversion. Because such a process often expels local residents who have depended on the nature around them, the exploitation of nature becomes irreversible.

In recent decades, de facto territorialization has occurred under the guise of "environmental protection," making it difficult to discern the government's true intentions. However, one thing is clear: These policies have allowed state power to penetrate into various domains and regions more deeply than ever before, often without anyone being aware of their impact. Let us observe how such a process unfolded in the Tonle Sap Lake in Cambodia.

4.2 Fishing and Politics in Cambodia

4.2.1 Tonle Sap, the Largest Freshwater Lake in Southeast Asia

The beginning of this chapter described the people living on and around Tonle Sap as fishers for half of the year and farmers for the remainder. A more nuanced view reveals that some individuals primarily depend on agriculture or fishing, while others alternate between the two based on the season (Sithirith 2016). The locals identify themselves as either land people (*neak leu*) or river people (*neak tonle*). Both groups produced goods essential for their survival, and through the exchange of fish and rice, they maintained a long-standing barter economy until more capitalistic forms of production became dominant. A society based on mutual dependence and multiple means of subsistence is a distinctive hallmark of the lake region.

However, the introduction and proliferation of cash-based exchange practices radically transformed the local economy away from the more common barter system. The seed of this pivotal shift was planted during the colonial era when the government established the fishing lot and began to enclose the fisheries on Tonle Sap Lake. Some fishers were forced to sell their catch at the market instead of bartering it for goods produced by their neighbors. Consequently, economic activity around the lake became dominated by market trends and the interests of middlemen, while the needs of local villagers were marginalized. Interdependence between land- and lake-based farmers collapsed, giving rise to new patron-client relationships. Fishers who succeeded in securing productive fishing lots expanded their businesses by employing and subjugating those less fortunate.

To understand the specifics of the fishing lot system, we must closely examine the geography of Tonle Sap Lake. Tonle Sap is divided into three distinct areas: fishing lots, common fishing areas, and protected areas. Until their dismantling in 2012, the fishing lots were subjected to strict supervision to prevent poaching. What makes Tonle Sap noteworthy is that the size and location of these supposedly distinct areas change as the lake floods. Tonle Sap expands dramatically during the rainy season, as shown in figure 4.2, and this annual transformation necessitates a complex system of resource management.

The start of the rainy season (between the end of June and the beginning of July) causes the Tonle Sap River to reverse its course, flowing toward the lake rather than away from it. This reversal continues until mid-October, during which time the lake swells to five times its regular size (marked as "inundation extent" in figure 4.2). As the lake expands, its average water depth increases from 1–2 meters to 8–10 meters (Kasai 2003, 43). Tonle Sap is surrounded by its famous flooded forests—swampy wetlands that are regularly submerged.

Rapid economic development in the Mekong region and the expanding tourism industry around Tonle Sap Lake in recent years have disrupted this remarkable ecosystem. The construction of large hydropower dams in upstream Laos and on the tributaries in the Lower Mekong Basin is altering the flow regime, particularly affecting the Tonle Sap Lake and the Mekong River (Yu et al. 2019). Several international organizations, including the Asian Development Bank (ADB), the UN Food and Agriculture Organization (FAO), and the Mekong River Commission (MRC), have formed a coalition to protect the lake's environment and prevent further damage. Tonle Sap Lake is increasingly viewed as a "global space" for biodiversity conservation. However, regional and national actors are shaping different approaches and systems to influence its governance based on various interests and political motivations, complicating the management of water and resources in the lake (Sithirith 2022).

According to Sokhem and Sunada (2006), the core problem for the country's fishing sector is the weak enforcement of laws and regulations. Research by Degen and Ratner indicates that the dwindling catches of smaller-scale Tonle Sap fishers are closely correlated with the overall decline of fish stocks in the lake and that increasing competition exacerbates the issue (Sok et al. 2022; Ratner et al. 2011; Degen et al. 2000).

Ratner (2006) suggested that a system of community fisheries could help mitigate the emerging ecological crisis. This system would involve creating fishery cooperatives with established operational rules, as well as recognizing fishers' rights to access and utilize specific areas. This approach is similar to what the Cambodian

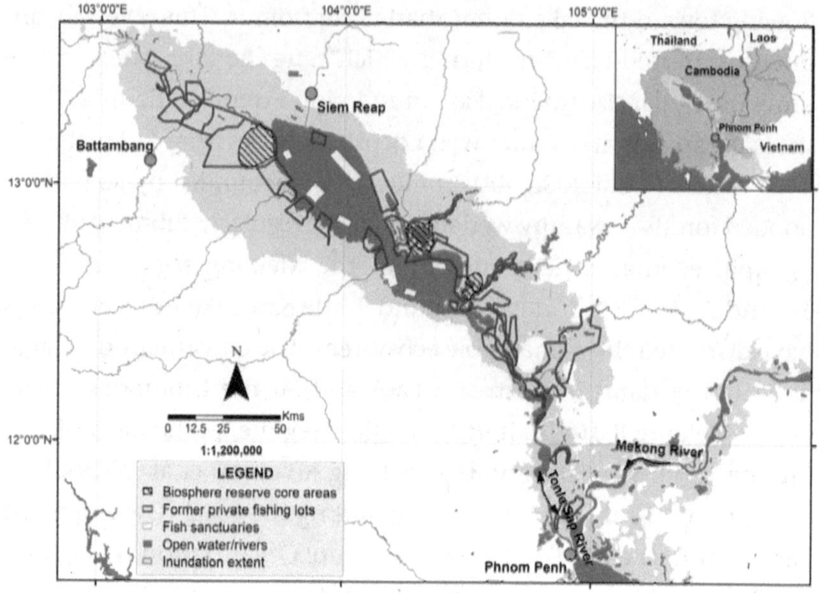

Figure 4.2
The flooding of Tonle Sap Lake and the locations of fishing lots. *Source*: Cooperman et al. (2012, 103).

government implemented in 2012 when it transferred control of the Tonle Sap fisheries to local communities to reduce illegal fishing. Then-Prime Minister Hun Sen stated in his July 1 (2012) address, "Electrocution (deemed illegal) is a very dangerous way of catching fish and will lead to the depletion of our aquatic animal resources. I am begging our people not to use this kind of gear to catch aquatic animals" ("Prime Minister Outlines Cooperation Measures to Protect Fisheries Resources" 2012, 31).

As seen above, government intervention was aimed primarily at curtailing conflict, but it was also an attempt to promote conservation. The government converted commercial fishing lots into research and conservation sites, and core areas were established for biodiversity conservation, with the Ministry of Environment assigned to oversee management and protection (figure 4.3). Despite the state's claim

Figure 4.3
A sign warns fishers not to enter the conservation zone. *Source*: Photo by Gerald Flynn for Mongabay).

that this move aimed to foster decentralization and democratization, many NGOs at the time doubted that the strategy would achieve its intended effect, given the local communities' lack of capacity to enforce such policies.

Are the community fisheries governing Tonle Sap truly benefiting local people in a way that guarantees sustainable resource use? One assessment was conducted in the Ek Phnom district in Battambang Province on Tonle Sap between 2014 and 2015 (Vong 2019). It found that the lack of enforcement of communal rights failed to exclude outsiders. Additionally, it revealed the ineffective development of alternative income sources after conservation measures were tightened. Such empirical studies, based on extensive fieldwork, offer significant contributions to understanding the realities behind the seemingly benign "community-based" practices. However, they overlook the critical role that lot owners play as mediators among fishers, business entities, and the state. The following section addresses this limitation by using information gathered during my interviews with former fishing lot owners to explain the changes resulting from the abolition of the fishing lot system. I also highlight the evolving relationships between the government and local communities around Tonle Sap.

4.2.2 The Origins of the Fishing Lot System

Historical context is essential for understanding the significance of the fishing lot cancellation. Tonle Sap's fishing lot system dates back to the nineteenth century when the government sought to commercialize fisheries on the lake and established a system to enclose sections of it. At that time, Cambodia was still a French protectorate, and marine products constituted the bulk of the country's exports. Eighty percent of these products were salted and dried fish, primarily catfish caught in Tonle Sap (Kikuchi 1981, 501). Adhémard Leclère, a key figure in the French administration near the end of the nineteenth century, noted that during the reign of King Ang Duong (1840–1860), fishing rights to specific locations were usually granted for free. This changed when King Norodom (1860–1904) began financially leasing fishing rights to fund the construction of a royal palace in Phnom Penh (Kikuchi 1981). However, fishing along the shores of Tonle Sap remained free for everyone at that time. King Norodom accumulated much of his wealth by selling monopoly rights to Chinese entrepreneurs, who were his chief trading partners (Cooke 2011).

It is difficult to pinpoint a specific date for when the fishing lot system in Tonle Sap took shape. Most researchers assume that the system was created by French colonial administrators when they implemented the first fishing lot–related policies around 1908. However, the reality may be more complex. Kikuchi demonstrates that the leasing system for fishing rights, introduced by King Norodom for small and mid-sized rivers and lakes (excluding Tonle Sap), gradually expanded into the surrounding areas (Kikuchi 1981). French rule may have simply facilitated the system's encroachment on Tonle Sap Lake.

France codified the fishing lot system in Tonle Sap in 1908 to streamline tax collection efforts. Fish harvests fluctuated from year to year, making the product difficult to tax. Taxing fishing rights proved to be much easier, generating a steady income for the government's coffers. The new French system also aimed to limit abuses by indigenous tax collectors who exploited their middleman position to take advantage of fishers. The amendments established a system in which fishing lots were auctioned off to a relatively small group of owners, who then paid taxes directly to the state. From 1900 to 1920, annual revenue from the fisheries sector (including income from export taxes) accounted for around 10 percent of Cambodia's national budget (National Archive of Cambodia, file no. 24105).

The system developed during the French Indochina era became the foundation of today's resource management system in Tonle Sap. The tax rates were based not only on the fishing lot itself but also on the type of equipment used. Fishing lots were classified and taxed according to the scale of fishing operations: large, medium, or small. Fishing lots for large fisheries—which utilized enormous stationary trawl nets—were reallocated through a competitive bidding process every two years. Medium-size fisheries require a license and prior approval to use specific types and quantities of fishing equipment.[3]

A year after the introduction of the fishing lot system, the fishing industry contributed one-ninth of the government's entire budget, a significant increase. Nevertheless, illegal fishing remained rampant, and policies aimed at conserving fishing resources proved ineffective (Cooke 2011). By the 1930s, France had begun to implement extensive conservation policies, leading to a flurry of laws and regulations.

After gaining independence in 1953, Cambodia retained the old French fishing lot system. The postcolonial government established the Department of Fisheries (later renamed the Fisheries Administration) in 1965 to enhance fishery resource governance. While the

Fisheries Law was overhauled, these changes were relatively superficial, and the underlying system of designated fishing lots remained intact. As one fisher who worked in Tonle Sap from the 1960s recalls,

> When I came of age in the 1960s, we had a fishing lot system and protected areas, and everyone took the boundaries of these areas very seriously. Lot owners operated only in their designated areas and never encroached upon the lots of others. The protected areas were also strictly enforced, and local fishers respected their boundaries. If lot owners strayed beyond the borders of their territories and attempted to fish in public fishing grounds, it would cause an uproar. (interview with a villager in the Kampong Phluk Commune, Prasat Bakong District, Siem Reap Province, October 2012)

This account underscores that people were more concerned about lot owners intruding into public fishing grounds than about small-scale poachers in the fishing lots.

Between 1970 and 1979, the rise and subsequent rule of the Khmer Rouge triggered a civil war that greatly affected the Tonle Sap area, plunging the fishing lot system into chaos. Pol Pot's regime forbade fishing altogether and forced people to work in the rice fields in cooperative units based on communist ideals. While some Khmer Rouge cadres may have continued fishing in Kompong Chhnang Province, we know very little about events in the lake area during this period. Certainly, there was virtually no commercial fishing on Tonle Sap for ten years, and by the time operations resumed in the 1980s, the fishing grounds were richer than ever.

4.2.3 The Ramifications of the Fishing Lot System

The reintroduction of the fishing lot system in 1987 marked a turning point in the territorialization of Tonle Sap (figure 4.4.). In theory, fishing lots were to be auctioned off to the highest bidder at regular intervals, but in reality, the same fishing lots were repeatedly claimed by a small group of politically well-connected fishers. Fishing lot operators were required to adhere to regulations outlined in the

Figure 4.4
Fishers with small nets working on Tonle Sap Lake. *Source:* Photograph taken by Nobuyuki Yagi (March 2015).

so-called "burden book," the certificate that accompanied the fishing lot. This certificate included important information such as a map indicating the lot's location, rules to ensure proper management and conservation of fish, and the amount paid for the lot. However, the actual amounts paid were often much higher than those stated in the burden book; one fishing lot owner claimed to have paid almost ten times the amount indicated on the certificate. Fishers rarely followed the certificate guidelines. For example, while subdivision of fishing lots was prohibited, the overwhelming majority ended up being exploited by multiple extralegal subcontractors. Furthermore, local officials and politicians often demanded various perks in exchange for granting exclusive fishing rights to lot owners.

To recap, the fishing lot system functioned for about a century, except for a brief interruption during the Khmer Rouge communist

Table 4.1
Number of fishing lots, average size per lot, and number of fishing lot-related conflicts

Year	No. of fishing lots	Total area of lots (ha)	Average area per lot (ha)	No. of conflicts
1998	164	390,000	2,378	826
1999	155	953,740	6,153	1,990
2000	83	422,203	5,086	1,258
2001	82	422,203	5,148	493

Source: Ishikawa et al. 2008, 319.

regime in the 1970s. Over time, the total area allocated to fishing lots gradually decreased, as shown in table 4.1. This reduction was the result of government-mandated opening of fishing grounds, which can be referred to as "de-territorialization." De-territorialization occurs when a government relinquishes its control over previously enclosed resources. For instance, in Cambodia, the total area of fishing lots in 1919 was 1,434,710 hectares. By 1998, this had dwindled to 390,000 hectares, representing less than a third of the original area. However, while the total area of fishing lots continued to decline, the size of individual fishing lots remained stable or even expanded, as indicated in table 4.1. This suggests that access rights to fishing resources became concentrated in the hands of fewer lot owners, who evolved into either employers or subcontractors.

Table 4.2 shows that the total area of fishing lots increased by 100,000 hectares between 1998 and 2000. This period was marked by numerous—sometimes violent—conflicts between lot owners and small-scale local fishers. Frequent disputes compelled stakeholders to petition for government mediation, resulting in an increase in government interventions.

A significant intervention by the Cambodian government in the Tonle Sap system occurred in March 2000 when then-Prime Minister Hun Sen announced that he had instructed the Cambodian Ministry of Agriculture, Forestry, and Fisheries to develop a plan to reduce the number of fishing lots and convert some of these into open-access public fishing grounds. This policy resulted in a 56 percent reduction in the total area of fishing lots on the lake. By 2012, as mentioned earlier, the system was abolished entirely. An official from the Fisheries Administration stated that the government took over half of the area that had been essentially privatized as fishing lots, designating it as an open-access fishing ground to be managed in cooperation with local communities.[4] In practice, 76.37 percent of the total area previously occupied by fishing lots was returned to the community, while the remaining 23.63 percent became state-designated conservation areas. The conservation area became the main battlefield where illegal fishing by outsiders went out of control.

There is little statistical evidence regarding the ripple effects caused by policies aimed at opening fishing lots in Cambodia. One available source is the Fisheries Administration data on known conflicts occurring in the years following the partial opening of fishing lots in 1998 (see table 4.2). While the interpretation of these figures is not straightforward, they confirm that the number of disputes skyrocketed in the late 1990s. In other words, although the government reduced the absolute number of fishing lots and increased the total area of public fishing grounds, this was accompanied by a rise in disputes. Why might this be? One possibility is that fishers from different areas clashed over access to the newly opened fishing grounds. It is also possible that some fishing lot owners, who were ordered to open their lots, refused to comply.

Although my interviews conducted in January 2018 with Fisheries Administration officials did not clarify the exact number of disputes, they did provide information on the types of conflict: disputes over illegal fishing, dredging of sand in protected areas, deforestation, tourism development, and the construction of hydroelectric power plants—conflicts generated by actions and measures related to economic development. Given that Cambodia is experiencing rapid

Table 4.2
Changes in the size of fishing lots (ha)

Province	1919	1940	1998	2000	2001	2001–2012
Kampong Chhnang	67,667	63,037	NA	62,256	45,084	17,172
Kampong Thom	248,272	192,571	NA	127,126	69,353	57,773
Siem Reap	NA	NA	NA	83,941	22,725	61,216
Pursat	105	NA	NA	55,120	24,848	30,272
Banteay Meanchey	182,352	189,362	NA	332,756	6,411	26,358
Battambang	NA	NA	NA	146,532	102,718	43,814
Total	1,434,710	444,970	390,000	507,731	271,139	236,605

Source: Compiled by author from Vikrom and Sithirith (2008) and FiA (2012).

growth, these conflicts are likely to increase, even if territorial confrontations over access to fishing lots eventually subside.

4.3 Why Did the Government Relinquish Control over the Fishing Lots?

Several hypotheses arise regarding why the Cambodian government decided to open the fishing lots. One factor is the changing administrative incentives of the state. The likely existence of oil and gas reserves beneath Tonle Sap Lake prompted the government to conduct further studies on exploitation possibilities.[5] For the government, the first step in securing control over the area was to dismantle the system that granted rights to private entities over parts of the lake and establish itself as the main beneficiary of future revenues from the exploitation of these subterranean resources. In essence, the emergence of the environmental state has continued to uphold the "developmental core," not necessarily by pursuing growth and wealth but by reinforcing centralized governance and control. This process has been wrapped in the logic of the environmental state, including principles such as conservation, which often obscures the political effects of such interventions.

Aside from uncertain benefits, a more concrete advantage of the abolition scheme accrued to small-scale fishers whose activities were not taxed by the government. From a revenue perspective, it may have been better for the government to leave the fishing lots in place and maintain the associated revenue streams. If so, why did the Cambodian government intervene in Tonle Sap in a manner that seemingly harmed its revenue? My fieldwork suggests that the state intended to utilize the lake's natural resources for redistributing economic profit and ensuring political stability. In other words, the regime hoped to gain support from a broader segment of the population surrounding the lake by redistributing fishery resources that had previously been dominated by a limited number of lot owners. This raises the possibility that, instead of attracting votes through subsidies, infrastructure projects, and other economic incentives, the government aimed to achieve its political objectives by promoting larger goals, such as reducing inequality and protecting vital ecosystems.

Can this agenda explain the situation around Tonle Sap? As of 2015, fishing—which includes industrial fishing, household fishing, and open-field fishing (for example, in rice paddies)—contributed about 8 percent to Cambodia's GDP (MAFF 2015). According to Hun Sen's statement on March 8, 2012, approximately one hundred fishery companies were involved in industrial-sized operations, and this small cluster generated about $400 million in total revenue. The government's new interventions in the fisheries sector aimed to distribute these revenues more broadly. In an extensive speech on March 22, 2012, the prime minister spent nearly three hours explaining the fishery reform. Starting with the historical development of fisheries and the fishing lots system, he then outlined the main reasons for state intervention in Tonle Sap: to reduce conflict between private lot owners and fishers, to conserve the lake's natural resources, and to reduce poverty levels in the surrounding communities ("Prime Minister Outlines Cooperation Measures to Protect Fisheries Resources" 2012). He also criticized the Fisheries Administration for the sector's many problems. According to the prime minister, this fishery reform was a historical event that only his regime had been capable of undertaking.[6]

The intent to redistribute fishery profits, upon inspection, proves insufficient as an explanation. This intervention must be examined against the backdrop of a wave of decentralization that had dominated Cambodian politics for a decade. I believe that state interventions in Tonle Sap were linked to the emergence of democratic governance heralded by the country's general election in 1993 (Öjendal and Lilja 2009; Peou 2007). Around the turn of the millennium, Cambodia was undergoing a decentralization process accelerated by new legislation mandating communal elections and the enactment of the Law on Commune/Sangkat Administrative Management in 2001. Until that point, municipal leaders had occupied seats in local government. Starting in 2002, for the first time, Cambodians across the country directly elected their community representatives (Slocomb 2004; Mansfield and McLeod 2004; Öjendal and Sedara 2011). Further, local elections under the new laws were held in 2007 and 2012, resulting in victories for the ruling Cambodian People's Party (CPP). The first commune council elections in 2002 were not only a milestone for the country's democratization process but also a symbol of decentralization.

Commune councils create a platform for various political party members to debate and decide on local policy matters. By the 2010s, the power of the CPP in these councils was absolute and unchallenged, despite occasional "reforms" of the electoral system. Therefore, if the ruling party already enjoyed such a solid power base, why did it continue to intervene in Tonle Sap's fisheries arrangements?

In 2011, Cambodia had a population of 14.5 million, over 4 million of whom—one in four—had a direct or indirect stake in fishing on the lake (Sithirith 2011). The government was anxious to manage politically sensitive conflicts between small-scale fishers, who comprised the vast majority of these four million stakeholders, and a smaller group of fishing lot owners, numbered in the dozens. While the issue was relatively minor in terms of tax revenue, it had potentially dramatic political ramifications. The general election of July 18, 2013, resulted in a narrow win for the CPP, while the opposition Cambodia National Rescue Party significantly increased its seats in the National Assembly. Allegations of fraud by the ruling party rocked the country's still-fragile democratic foundations. Following the elections, Prime Minister Hun Sen orchestrated the dissolution of the Cambodia National Rescue Party. The next general election, in 2018, ended in a decisive victory for the CPP among twenty political parties.

As mentioned earlier, the Fisheries Administration benefits from its connections with fishing lot owners. For instance, fishing lot owners often provided accommodation to politicians and officials during their visits to local areas. However, such connections raised the risk of perceptions of corruption. Therefore, it made sense for the government to address escalating conflicts between fishing lot owners and small-scale fishers before discontent spread throughout Cambodia. In the post-2000 period, growing resentment among small-scale fishers frequently led to minor conflicts across the country. The government's decision to initiate a full-scale investigation in 2011 was viewed as an expression of Prime Minister Hun Sen's frustration with the Fisheries Administration's inability to resolve the situation.

The issues surrounding Tonle Sap involve multiple government agencies, primarily the Ministry of Agriculture, Forestry, and Fisheries, as well as the Ministry of Environment and the Ministry of Water Resources and Meteorology. The Ministry of Agriculture, Forestry, and Fisheries has jurisdiction over fisheries in the fishing lots. The Ministry of Environment oversees protected areas, primarily focused on biodiversity preservation. Established in 1999, the Ministry of Water Resources and Meteorology includes the Tonle Sap Authority, whose chairman is reportedly a confidante of the influential former Prime Minister Hun Sen. Notably, while the Fisheries Administration—which operates under the Ministry of Agriculture, Forestry, and Fisheries—technically has authority to address illegal fishing operations, this responsibility is effectively managed by the Tonle Sap Authority. Further research is necessary to clarify how the balance of power among these various agencies has influenced the abolition of the fishing lot system. One thing is clear: The Ministry of Agriculture, Forestry, and Fisheries lacks the political resources to maintain the rights associated with its jurisdiction over the fishing lot system.

Many fishing lots have been designated as communal fishing grounds and protected areas. However, this raises another question:

Who will manage these areas and how? Most community fishery groups have incentives to govern their fishing grounds responsibly. It remains to be seen whether they will be able to do so effectively in practice without sufficient resources for implementation.

It is also uncertain how effectively the Ministry of Environment will conserve the new "protected areas" under its jurisdiction. The policy of opening up fishing grounds was met with optimism by small-scale fishers.[7] However, some are concerned that Tonle Sap Lake is on the brink of becoming a "tragedy of the commons." A report based on more than five hundred interviews with fishers indicated a significant decline in both the variety and volume of catches, even in the newly established "conservation areas" that have received support from the European Union (Bahadur et al. 2020; Jones 2015). Furthermore, there are obvious limitations to what communities can achieve in resource management, particularly regarding the control of illegal fishing activities. This issue will be discussed in the next section.

4.4 How "Returning Power to Local Communities" Can Trigger Inversion

Historically, the fishing lot system provided a private means of managing fish that exhibited many characteristics of a commons. From this perspective, I believe the government dismantled a century-old system of private governance and replaced it with a policy of public governance by local communities. This approach is the opposite of the more familiar governmental strategies of privatizing or nationalizing resources that were once common to locals.

Overall, the reaction of small-scale fishers to the complete opening of the Tonle Sap fishing grounds has been positive. Previously, fishing lots were fenced off or patrolled by armed guards who reacted violently if fishers entered areas outside their own. While those restrictions have been lifted, challenges remain. Representatives from local

NGOs express concern about the increasing incidents of overfishing and the use of illegal methods, such as electrofishing. In agreement with the government, each community fishery association has their own designated zones for fishing practices and conservation according to the plan. Such boundaries are indicated on the map and sometimes by physical markers such as signposts. Despite clear boundaries with GPS technologies, most communities are ill-equipped to address these issues, lacking even the funds for minor expenses like the gasoline needed for patrolling community domains for poachers from outside and the use of illegal fishing tools. Following the decentralization policy, reported instances of illegal fishing and conflicts among fishers have decreased in certain areas (Bahadur et al. 2020).

There are ongoing allegations that officials can be bribed to overlook overfishing. Sithirith (2014) suggests that officials are susceptible to kickbacks due to their low salaries. The opening of fishing lots has done little to eliminate these corrupt practices; instead, it has made them more complex. The ambiguity surrounding what constitutes "illegal fishing" created a breeding ground for corruption. Illegal fishing includes poaching in another community's designated waters, fishing in the conservation zones, fishing in the closed season, and using prohibited tools and nets. Illegal fishing operations continue to evade the law through well-placed bribes to police and Environmental Conservation Bureau officials. Such corruption forced certain fishers into corruption themselves because continuing to fish by whatever means, including paying bribes, is the only way for them to survive (Flynn and Srey 2022).

In short, the opening of the fishing lots has created ambiguity that now poses an obstacle to the government's stated goals of reducing inequality and preserving resources. Tonle Sap was subjected to privatization for many years, operating specifically under the logic of exclusion, and later, power was delegated to communities—a process distinct from nationalization. As a result, although not uniformly across all areas of Tonle Sap, disputes among villagers have mounted as boundaries blur and patrols are inadequate. This situation is exacerbated by the accelerated rate of resource depletion. The unforeseen outcomes of the opening-up policy suggest that the longstanding fishing lot system had some fragile merits. The tragedy of the commons is now becoming a reality on Tonle Sap Lake.

The hailed "release" of resource access under the guise of community empowerment should be scrutinized to uncover the state's hidden aims. I argue that the Cambodian government is willing to intervene in Tonle Sap because millions of people *depend* on the lake's fish for their livelihoods. Given that Tonle Sap remains relatively marginal in terms of overall tax revenue and economic production, implementing a popular fisheries policy on the lake represents an easy way for the state to appease at least a quarter of its citizens without much financial investment. It is no coincidence that Hun Sen's government introduced its successive fishing lot policies just before elections. The state undoubtedly recognized the political value of interventions in the lake's fisheries, whether through enclosure or opening. Officials continue to obscure possible state motives behind claims of decentralization and democratization. Moreover, the real social and environmental effects of these populist tactics remain unexamined. If approaches like those taken by the Cambodian government are indeed deliberate tools of governance, they warrant scrutiny (Dina and Sato 2014). Development objectives embodied in the environmental policies are in fact the key features of an environmental state.

The traditional fishing lot system was exclusionary, as it represented a private management of the commons. Nonetheless, it remained stable for extended periods, underpinned by a sense of order among fishers on the lake. Confusion and the reconfiguration of corruption may be inevitable when such a system is dismantled and resources are opened to the management of local communities that vary not only in size but also in their capacity to govern. The Fisheries Administration, the government department in charge, seems incapable of restoring order around Tonle Sap. Consequently, other groups, such as NGOs, have become a stopgap until the policy issues are resolved. Hardin's theoretical predictions echo today, half a century after he made the statement, "Injustice is preferable to total ruin."

This chapter demonstrates that even when the economic importance of natural resources diminishes, these resources can still serve as a powerful political tool if a significant portion of the population relies on them. For governments, the option to manipulate access to such resources may hold greater value than direct economic benefits derived from imposing taxes or providing subsidies. This is particularly true in countries where many people depend on primary industries for their livelihoods. Those living near natural resources often have less political power due to their marginal positions in the rural periphery, making them less aware of unfair distribution when access to resources is controlled at the central state level. Once a government recognizes these dynamics, the temptation to manipulate resource access for political gain increases.

In 2016, the environmental foundation Global Nature Fund declared Tonle Sap "the most threatened lake in the world," highlighting the decline of the natural environment and the depletion of resources impacting this vast region (Seiff 2017). Although we are not in a position to examine what the conservation outcome would have been if there had been no fishery reform, distrust of local conservation officers is now widespread, incentivizing both fishers and officers to engage in bribery, which then invites inversion. Fishers now have less incentive to cooperate with the state in conserving the lake in the long term. A lack of trust for local authorities is widespread due to the democratization of resource access. A recent report by a group of scientists revealed that the issues extend beyond the deterioration of water availability and access to fishery resources; they also encompass health risks posed by heavy metals, pesticides, and microbial pollution (Shivakoti et al. 2020). Additionally, the risks associated with hydropower developments upstream on the Mekong River further threaten the ecology of Tonle Sap, contributing to declining fish catches in the region (MRC 2022).

New policies aimed at empowering local communities and protecting the environment have produced contradictory effects in some areas of Tonle Sap. Delegating governance to communities can lead to inversion when responsibilities are unclear and the affected communities lack the necessary resources for effective management. Simply handing over resources for conservation cannot prevent these issues. Incentives to utilize the resources must be accompanied by social and organizational skills to sustain the flow of benefits, ensuring that each stakeholder can demonstrate fair commitments.

Faced with a more recent enforcement action by the government in March 2022 aimed at cracking down on "illegal fishing," to achieve conservation in the lake, one fisher commented, "By continuing to fish, we are forced into hiding; we are forced into crime" (Esterman 2022). Signs of inversion reappeared as patterns of exclusion continued to drive away potentially the most effective stewards of the lake—small-scale fishers—from their resources. Though fishing serves as the people's livelihood, the drastic change in resource conditions requires state intervention, making the rise of the environmental state inevitable. A lack of trust in the government is undermining these potential stewards, preventing the environmental state from achieving its conservation goals.

It will take considerable time for the Cambodian government and its people to collaborate on managing the lake's abundant resources in a way that both addresses inequality and protects the ecosystem for the future. Neither government reforms nor local initiatives alone will suffice to resolve these challenges. Ultimately, trust among the key actors, shaped by their dependence on the resources each actor offers, will determine the lake's future.

5 Compulsion: Maintaining Irrigation Infrastructures in Indonesia

In this chapter, I examine the case of state-constructed irrigation works in Indonesia to explore the inversion of state-led water resource development. Water resource development is one of the oldest forms of large-scale state intervention in nature. Irrigation infrastructure not only transports water but also impacts the underlying state-society relations, inviting either cooperation or resistance from local communities. The chapter focuses on the maintenance work required for irrigation facilities to function and investigates how this maintenance further affects the relationship between states and local societies. The implications of the maintenance dimension extend far beyond what the original designers of the irrigation systems anticipated. Through this analysis, the chapter delves into the ongoing process of inversion.

5.1 Maintenance as an Intersection of State and Society

Aqueducts, which ensure a stable water supply throughout the year, are the oldest known examples of social infrastructure in Indonesia. Irrigation facilitates flourishing agriculture while also creating pathways for the transport of goods and people. Moreover, it plays a vital role in flood prevention. Across the rain-heavy regions of Southeast Asia, irrigation has significantly improved community livelihoods. However, we also observe a history of failure in the state's attempts to increase production through irrigation development. A notorious example is the Mega Rice Project, initiated in Indonesia in the mid-1990s to achieve self-sufficiency in rice. It has been reported that a vast canal system, along with other interventions in Kalimantan, has destroyed forests, triggered massive forest fires, and marginalized local populations (Boehm and Junaid 2015). Poor resource management ultimately led to the expulsion of the very people the project needed to succeed in self-sufficiency of rice, thus constituting a typical example of inversion.[1]

I suggest we zoom in on irrigation as a method to utilize water resources. The universal laws of physics dictate that water flows downhill. This fundamental principle requires irrigation planners to consider both engineering challenges and the social dynamics of their projects. First and foremost, upstream residents experience different benefits and face different burdens than those living downstream. For instance, upstream communities primarily benefit from the water supplied to their fields, while downstream residents may be more concerned with irrigation's role in flood prevention. These differing priorities can lead to conflicts regarding the sharing of responsibilities for the construction and maintenance of the physical aqueducts.

Let us visualize the problem to better understand what is at stake. Imagine two hypothetical villages: village A, located upstream, and village B, situated downstream (figure 5.1). The government constructed a large arterial canal from which two smaller canals branch off toward the two settlements. Farmers then dug field channels from these branch canals to irrigate their fields. Village A, upstream, enjoys a more secure water supply, whereas village B, downstream, faces a precarious situation. Before the water reaches village B, it can be lost

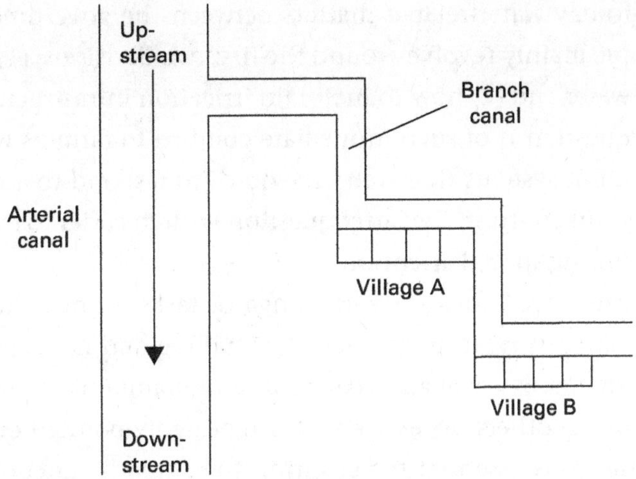

Figure 5.1
Typical configuration of agricultural irrigation infrastructure. *Source*: Created by the author.

in various ways: leaks, evaporation, increased water usage upstream due to a growing number of farmers, unforeseen failures in the water distribution system, and more (Freeman and Lowdermilk 1991).

A local community seeking to implement an irrigation system must make three important decisions (Coward 1980, 19): (1) how to deliver water to end-users, (2) how to maintain and manage the irrigation infrastructure, and (3) how to resolve conflicts between users. These decisions involve water supply issues and the distribution of burdens—political issues that expand the debate beyond civil engineering. In many Asian societies, the state has often played a significant role in maintaining infrastructure and resolving conflicts (Montgomery 1983, 94). The state has a vested interest in conflicts over irrigation between villages or regions, viewing these disputes as threats to production and, consequently, to the country's financial stability (Wolters 2007, 226). Conflicts over irrigation are typically significant enough to necessitate state arbitration; minor disagreements that villagers can resolve among themselves are comparatively rare.

Day-to-day water-related matters between the government and the people mainly revolve around the first two decisions: (1) how to deliver water and (2) how to maintain irrigation infrastructure. The former question is of such immediate concern to farmers who rely on irrigation systems that states are quick to respond to any issues that arise. In contrast, the latter question is often easier to overlook, thus requiring special attention.

"Maintenance" covers a wide range of tasks, from felling trees and weeding to creating grasslands, installing and managing flow meters, dredging waste and sediment, and maintaining pump stations, among others. As experts have repeatedly pointed out, such maintenance is essential for ensuring the smooth functioning of irrigation infrastructure (Sagardoy et al. 1982). A facility that falls into disrepair due to poor maintenance represents a colossal waste of investment. Less obviously, poor maintenance also results in significant economic losses. When farmers cannot rely on a consistent water supply, they become hesitant to cultivate high-yield varieties that require precise amounts of water at specific times (Freeman and Lowdermilk 1991).

In Asia, the primary demand for water comes from rice production. Rice is one of the most important staple foods in the world, with Asia serving as both the major consumer and producer (Redfern et al. 2012). Approximately 45 percent of the rice-growing area in Southeast Asia is irrigated, with the largest areas located in Indonesia, Vietnam, the Philippines, and Thailand (Mutert and Fairhurst 2002). For rice farmers, water shortages due to poor irrigation maintenance can quickly lead to life-or-death situations. When a drought destroys a rice crop, farmers cannot recover the investments made in raising seedlings, preparing fields, cutting grass, controlling pests, and more. In earlier periods, when tax revenue primarily came from land rents collected from citizens, a malfunctioning irrigation system could be catastrophic for the state. This explains why states have historically compelled locals to maintain state-constructed irrigation infrastructure.

Akio Takahashi, who studied dredging operations in the Kyaukse District of Myanmar in the early 1990s, described the situation on the ground as follows:

> When the main and branch canals still contain flowing water, the government's Irrigation and Water Utilization Management Department is responsible for repairs to damaged areas. However, when substantial repairs or dredging of accumulated mud in the canals are needed, the water flow into the canals must be halted. Once the canals have been emptied, significant labor investment is required to complete the necessary work. This massive effort, in which water gates are closed and all intake and distribution of water must temporarily stop, occurs every January during the dry season. From each household near the canal, one adult laborer is mobilized from all individuals over 18 (both male and female). Non-farming households are also required to provide labor, as all residents near the canals use the water for cooking, washing, bathing, and other daily activities. (Takahashi 1996, 188, author's translation)

This example illustrates that even large-scale irrigation works must be maintained and managed locally. The crucial challenge is organizing and mobilizing the local labor force. Scholars are currently debating whether the foundations of the Burmese state could only have been established in arid central Myanmar due to large-scale irrigation systems that could only be constructed and maintained through compulsory and forced labor (Taylor 2009, 44).

In this chapter, I focus on one aspect of maintaining irrigation infrastructure: how it serves to assert the state's power over local communities and how it can trigger inversion. Pioneering development expert Albert Hirschman was the first to highlight how certain development projects stimulate and eventually shape decision-making habits among people (Hirschman 1958). For instance, a lack of road maintenance leads to a gradual decline in the quality and quantity of output or to temporary damages with no immediate severe consequences. In contrast, a lack of maintenance in the airline sector can result in catastrophic outcomes, including significant loss of life and property. The implication is clear: Contrary to common belief, underdeveloped countries can achieve notable success in ventures involving complex technology. The technology of irrigation may be simple, but the methods of distributing water within the community are complex and can have serious consequences for agriculture.

Hirschman argues, "Underdeveloped countries may well make a surprising success of ventures with a complicated technology which *must* be maintained in top working order. It is in these industries that the maintenance habit can be acquired and then spread to the rest of the economy" (Hirschman 1958, 142). Once initiated, Hirschman observed, a development project inevitably necessitates upkeep. This engenders a new set of obligations related to whatever work is required to sustain the project, creating a "compulsion to maintain" (Hirschman 1958, 146). Hirschman's perspective does not center on the intentions or will of the individuals in charge; rather, it emphasizes the pressures arising from the geographical, material, and technological conditions that must be fulfilled for the infrastructure to function effectively. This framework of the compulsion to maintain suggests that the upkeep of development projects does not necessarily arise from a deliberate effort by the state to impose its will on unwilling locals. In fact, in some cases, local communities are motivated by the inherent demands of a development project to engage with the state and facilitate its maintenance.

In the case of irrigation works, the required maintenance tasks are profoundly influenced by the weather characteristics of rural environments, including seasonal changes and fluctuating rainfall. As Hirschman explains, "The penalty imposed and, hence, the compulsion exerted by nature are strongest when failure to complete the job by a certain date means that much or all of the work already accomplished is destroyed, swept away, or otherwise lost. Such a situation exists when a dam must be built up high enough to hold back the river once it is monsoon-swollen" (Hirschman 1967, 96).

The question then becomes, who should respond to the pressure exerted by this compulsion to maintain? Should it be the central or local government, or should the responsibility fall to the local people? This often remains unclear in the early stages of large construction projects, such as power plants or multipurpose dams, which typically suffer from a lack of consultation with local residents. In the 1980s and 1990s, numerous dams were constructed in Southeast Asia with the support of Japanese ODA (official development assistance) loans. These projects received significant criticism from NGOs and other civil entities, highlighting that the design process was often carried out without any public consultation or participation (Sumi 1988; Murai and ODA Chōsa kenkyūkai 1989; Suwa 1996). Poor communication frequently left all parties involved without a clear understanding of individual responsibilities, and the situation only deteriorated over time.

The following discussion focuses not on the process of constructing massive dams but on the state-society relations that emerge downstream from irrigation infrastructure projects as a result of their existence or operation. When it comes to the smaller offshoots of an irrigation system, specifically the field channels that branch from secondary canals, local government officials often leave it unclear who is responsible for maintenance. This ambiguity suggests that the endpoints of vast irrigation systems offer a valuable context for observing the daily interactions between government and farmers, as well as the forces that influence the process of inversion.

5.2 Irrigation Projects in Tropical Asia

5.2.1 State-Led Irrigation and the Particularities of Water

In tropical monsoon regions, water is abundant during the rainy season but often in short supply throughout the rest of the year. For local communities reliant on rice cultivation, nothing is more important than reliable access to water (Mushiake 2002). Rice has been a staple in Southeast Asia throughout history, and precolonial governments in the region consistently prioritized effective water resource management to prevent famines (Attwood 1987). Subsequently, colonial powers in Southeast Asia, which sought economic stability in their territories, invested heavily in constructing irrigation infrastructure in the regions under their influence, ranging from the Dutch East Indies (modern-day Indonesia) to the French-administered territories of Vietnam, Laos, and Cambodia (Booth 2013). These irrigation projects were deemed essential for increasing rice production, which was needed to feed local workers whose labor sustained the plantations.

At the start of the twentieth century, for instance, the colonial government of the Dutch East Indies implemented a new "Ethical Policy," ostensibly aimed at improving the welfare of the colony's people. One component of this Ethical Policy was the development of irrigation works to enhance rice production (Kato 2014). This initiative emerged from a broader need for colonial governments to justify their existence in light of growing criticism of their exploitative systems on humanitarian grounds. Colonizers sought to deflect such accusations by launching projects that benefited the subjugated peoples of the colonies. Additionally, they aimed to modernize the economies of their territories, which required increasing rice exports. However, the traditional irrigation systems, initially built to ensure the self-sufficiency of local communities on their limited lands, were ill-suited to these new objectives.

Postcolonial countries in Southeast Asia invited foreign capital investments and technical advisors from Western Europe to construct large-scale irrigation infrastructure (Furnivall 1956, 320). As historian Yoneo Ishii, a specialist in Thailand and its irrigation systems, noted, "[Large-scale] irrigation required materials and labor that went far beyond what villages or households could provide. Work had to be carried out across vast areas and managed on a national level" (Ishii 1978, 18–19). This immense mobilization of capital and Western technology distinguished postcolonial irrigation projects from traditional, small-scale initiatives. The impact of these projects was profound. From the mid-twentieth century onward, it became common for governments to use public investment to create irrigation infrastructure that was urgently needed to boost food production. Accompanying these initiatives were official departments responsible for irrigation and other bureaucratic systems that extended government control over water resources (Molle et al. 2011, 330). Southeast Asian countries that gained independence after World War II were all eager to develop massive state-run irrigation systems to enhance agricultural production, receiving active support from developed nations.

In summary, investment in irrigation systems did not cease following independence from former colonial powers. The transformation of agriculture and state-society relations concerning water continued to advance. Since the late 1960s, the growing economy has driven an increasing demand for food and electricity across Asia. This rise in demand is supported by robust economic expansion, population growth, and Southeast Asia's emergence as a global manufacturing and industrial hub (IEA 2024). To meet this growing demand, enormous dams have been constructed one after another, financed by the World Bank, the Asian Development Bank, and various streams of bilateral aid. There are 1,055 dam projects in the Mekong River region, including 661 operational, 54 under construction, and 331 planned (Ang et al. 2024). The operational dams are primarily located in Thailand and the downstream portion of China, while the proposed dams are predominantly in Laos, Cambodia, and the upstream portion of China. Irrigation infrastructure has become a symbol of a centralized, modern state and continues to represent that identity today.

US-led Western nations feared that public unrest stemming from population growth and rice shortages could leave Asian countries vulnerable to the threat of communism. To address this, massive foreign aid flowed into the region to establish the legitimacy of newly independent states and to combat poverty, which could attract communist ideologies (Barker and Molle 2004). Under these political conditions, Southeast Asian nations, where small-scale traditional irrigation techniques were the norm, experienced a sudden increase in "government-motivated, government-led, government-targeted projects" (Mase 1984, 163). These water management systems were vast in scale and primarily constructed in newly developing areas where farmers lacked the financial or technological resources to manage such systems independently. Ultimately, government-led irrigation projects made farmers more dependent on public institutions (Mase 1984, 164).

While raw material wealth typically flows from the periphery to the center through trade or taxation, water must be utilized in its immediate vicinity. As this chapter elucidates, the simple existence of irrigation infrastructure does not ensure that water reaches the farmers at the edges of the distribution system. The critical element is the presence of resident organizations responsible for maintaining and managing the infrastructure. Without them, central or even local government officials could never hope to keep up with the routine maintenance of the countless large and small waterways extending from the main canals into increasingly fine capillaries.

As I described in chapter 3, the exploitation of forest and mineral resources in peripheral areas required local labor. Wood and minerals were usually transported out of the regions from where they originated. Water resources are different; they are the lifeblood of the farmers who live nearby. Waterways also function as transportation arteries for the cities and rural areas they traverse, determining where people build their settlements. New settlements are established based on the water they extract, store, and distribute for various purposes, including transportation, drinking, and agriculture. These communities also develop technologies and social structures aligned with their use of the waterways (Abernethy 2011, 87). An important function of these social structures is to arbitrate conflicts over water and facilitate the sustainable use of this precious resource. For example, rural societies in Japan often constructed facilities called *bunsui* ("water distribution") to ensure that all water was fairly allocated. They also established the custom of *bansui* ("water rotation"), which delineated specific blocks of time during which particular users could access water (Sugiura 2008). In Thailand, examples can be found in the water-scarce northeastern region, where pressure-controlling valves are installed in the water pipe to retain the water so that upstream residents can access it. The existence and invention of these technological and social systems indicate that water disputes have plagued irrigated societies for a long time.

5.2.2 The Side Effects of Irrigation: Rising Demand and New Public Health Challenges

When the state's power penetrates deep into societies through irrigation, it expands laterally in unexpected ways. This often occurs along one or more of the following three circuits.

First, the increased water supply may stimulate additional demand.[2] Masahiro Nakashima, a longtime scholar of irrigation in Southeast Asia, illustrates how the supply of water often creates demand, rather than the other way around, using the example of dams. "Once a dam is built, farmers can get enough water to grow rice even in the dry season. Farmers then want to use more water in the hope of increasing their income. . . . In short, water use is not always the result of existing demand" (Nakashima 2003, 12, author's translation). Nakashima's example pertains to irrigation infrastructure built in areas that were traditionally reliant on rainwater, enabling farmers to grow additional crop cycles during the dry season. In this context, irrigation infrastructure is not constructed to satisfy existing demand; instead, demand is generated by new water management facilities. Additionally, new roads, initially necessary for transporting construction materials to the dam site, improve access to markets, further boosting demand for goods. This enables locals to adopt a new lifestyle, motivating them to increase their economic activity to sustain their newfound prosperity.

A second circuit involves health and hygiene, which might seem unrelated to water management. Many countries have discovered that irrigation infrastructure, while essential for national modernization, has one unwelcome side effect, particularly in rural areas. Irrigation ponds create ideal breeding grounds for mosquitoes, which can spread malaria (Baeza et al. 2011). These insects can reproduce as long as temperatures remain above 16 degrees Celsius, meaning that in tropical regions with warm winters, they can remain active year-round.[3] Malaria is a serious infectious disease that affects three hundred to five hundred million people annually, killing between one and three million (Sachs and Malaney 2002). Infections impose significant economic burdens on patients and their families. Addressing the mosquito infestation at the root of a local malaria outbreak may involve restricting people's freedom of movement and suspending various economic activities, including trade. Another contributing factor is the tendency of people in rural areas to clear forests and establish settlements where malaria-carrying mosquitoes are already present (Qadir et al. 2010). Ultimately, eradicating malaria necessitates that irrigation departments, which are responsible for designing and locating new reservoirs, collaborate closely with other authorities in education, public health, and hygiene. This cooperation allows irrigation to serve as a means for the state to extend its influence into seemingly unrelated areas.

The third circuit involves increased environmental degradation resulting from booming agricultural production. In recent years, farmers across Asia have resorted to excessive use of chemical fertilizers to enhance their production of cash crops. One concerning outcome is that water contaminated with these chemicals is often used for agriculture and other purposes without prior decontamination

(Qadir et al. 2010). Groundwater pollution and its associated health impacts are issues that have become linked to irrigation as development progresses. Addressing these new challenges requires a variety of administrative interventions, ranging from educating farmers about chemical fertilizers to assessing the quality of agricultural products and screening for any effects on human health. Consequently, irrigation in rural areas is connected to numerous problems that extend beyond simple agriculture and create intersections between the concerns of the state and the people.

5.3 The Aspects of State Power

5.3.1 Changing Roles of the State

A central question in the institutional theory of water concerns the roles that governments, local communities, private companies, and other actors play in the efficient distribution of water. Recall Karl Wittfogel's theories on hydraulic societies discussed in chapter 1. Wittfogel proposed that irrigation and agriculture can relate to each other in three patterns:

1. Rainwater agriculture, where farmers depend solely on natural precipitation.
2. Hydroagriculture, where farmers rely on water management through small-scale irrigation or local community efforts.
3. Hydraulic agriculture, where farmers depend on large-scale irrigation works controlled by a central government to prevent floods and expand agricultural production. This method typically requires massive construction projects (Wittfogel 1957).

Historically, rainwater agriculture and hydroagriculture were standard practices in most regions of Southeast Asia (Barker and Molle 2004). The widespread use of these methods left little room for the involvement of financially strained governments. Precolonial governments in Southeast Asian countries with monsoon climates lacked the capacity to prevent flooding through civil engineering techniques; their primary concern was how to *adapt* to cycles of drought and flood.

Traces of these adaptations can still be found in many countries. For instance, *muang fai,* a traditional small-scale irrigation system used in northern Thailand, is also found in Laos and China's Yunnan province (Ishii 1978; Tanabe 1981; Stott 1992). On the island of Bali in Indonesia, local communities continue to employ the traditional *subak* irrigation system—a community-driven water management method that has long been the subject of academic study (Geertz 1963; Lansing 2007). Community-led management of resources has often played a vital role in small communities that are heavily reliant on cooperation (Kusumastuti et al. 2021).

However, not all of Southeast Asia consists of tightly knit communities, nor are these traditional organizations always capable of adapting to change. The weakness of *muang fai* in Thailand, for example, has been identified as its inability to politically confront intruders and flexibly adapt to include new users with different demands and water needs (Charoenmuang 2006, 103). In fact, scholars have argued that many parts of Southeast Asia, including Thailand, are characterized by local communities bound by relatively weak social ties (Embree 1950; Mizuno 1981; Shigetomi 1996). This weakness may explain why state-led irrigation projects in these countries expanded so rapidly: Loosely connected communities often have fewer social resources to resist unwelcome development projects initiated by the government.

Against this backdrop, the state expanded its large irrigation projects to boost agricultural production. Figure 5.2 illustrates how the acreage of irrigated land in various Southeast Asian countries has expanded since 1961. In Asia, the proportion of areas equipped for irrigation was 20 percent of the cultivated area in 1961, the highest

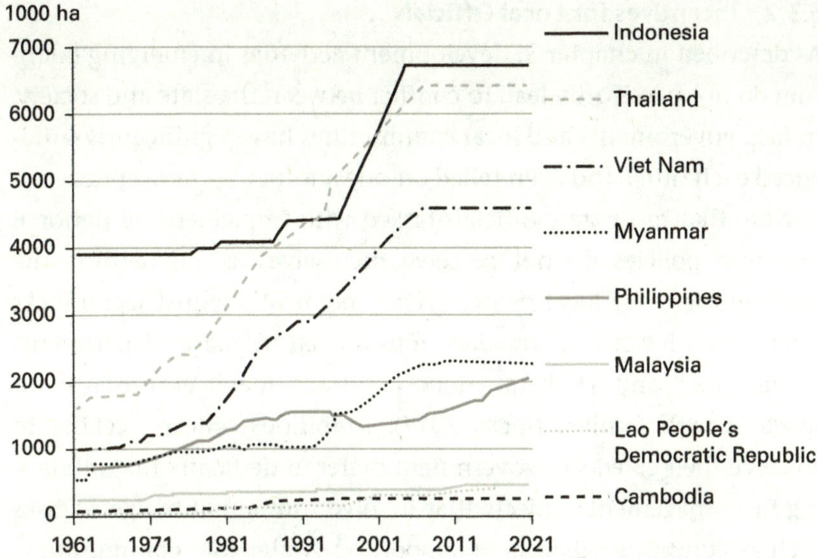

Figure 5.2
Increase in land area equipped for irrigation in Southeast Asia. *Source*: Based on data from the Food and Agriculture Organization (FAO) of the United Nations (http://faostat.fao.org).

among world regions (the world average was 10.2 percent). This figure increased to 39 percent in 2006, reflecting a rise of 19 percentage points, while the global average rose only to 19.7 percent—an increase of 9.5 percentage points (FAO 2011).

The total area of irrigable land in Asia amounts to 2.1 billion hectares. Of this, 1.8 billion hectares are currently being irrigated (Siebert et al. 2013). By the mid-1990s, most middle-income countries in Southeast Asia had irrigated all of their easily accessible land. In 2021, Asia had nearly 250 million hectares of cropland equipped for irrigation, which represents 42 percent of the total cropland in the region and 70 percent of the global cropland equipped for irrigation (FAO 2023, 7). Constructing new large-scale irrigation facilities incurs high marginal costs, resulting in few new constructions being undertaken. Most of the work on irrigation infrastructure today focuses on maintaining existing facilities.

5.3.2 Incentives for Local Officials

As described in chapter 3, development activities in emerging countries do not necessarily lead to conflict between the state and society. In fact, governments and local communities have significantly influenced each other and even relied on one another in these endeavors.

Note that most state officials tasked with implementing national irrigation policies do not perceive themselves as augmenting the state's power over local people. When political scientist Jacob Ricks interviewed lower-level officials from official irrigation departments in Indonesia and Thailand, none expressed much enthusiasm for working with farmers (Ricks 2017). Ambitious officers seeking to advance their careers in government prefer undertaking major building or management projects that involve substantial budgets. Tasks such as educating villagers or supporting "water user communities" are far less appealing. In a study of farmer water user communities in Cambodia, Mak Sithirith revealed how citizen organizations can falter when they receive only minimal support from a limited number of unmotivated government officials (Sithirith 2017). Farmers themselves often resort to bribery and deceit to keep officials at a distance in an effort to protect their holdings (Freeman and Lowdermilk 1991).

In summary, states seeking to enhance their water management face several challenges. The state relies on villagers to maintain irrigation facilities; however, officials responsible for implementing state policies have little incentive to build mutually interdependent relationships with these villagers. Conversely, villagers often have agency within their farms, households, and local communities, making them less susceptible to manipulation for broader public goals. What motivates them to undertake maintenance duties on irrigation systems is their daily reliance on the proper functioning of these systems in their geographic area. Therefore, the design challenge for irrigation systems is to address the incentive mismatch between the state and the villagers.

5.4 Dimensions of State Involvement

5.4.1 The Wide Reach of Water Resource Management

With significant disparities in incentives, what avenues exist for the state to penetrate local communities? How does the state maintain its influence? Under what circumstances does it choose to withdraw and delegate power to companies or other nonstate actors?

When we examine the variety of activities involved in irrigation in light of Wittfogel's analysis, it becomes evident that irrigation is not simply "productive" or "preventive." Irrigation also entails "preparatory" work, which includes civil engineering, labor, social organization, and compensation to residents. Additionally, it involves maintenance and management tasks that encompass not only the upkeep of infrastructure but also designing systems for water distribution. Importantly, irrigation incorporates "governance" work, a category that ranges from resolving conflicts over ownership or interests to planning for disaster management (such as drafting evacuation plans and establishing codes of conduct in emergencies). Let us take a closer look at these diverse tasks (table 5.1).

State interventions predominantly manifest as "productive" work, such as the construction of new irrigation canals. Local communities often welcome these interventions, as they typically align with community goals, such as increasing agricultural output or improving transportation access. However, it is crucial to note that the construction of large irrigation infrastructure is not universally embraced. Such projects often necessitate forced evictions that can cause significant temporary distress to residents. This was the case with several large irrigation works constructed with Japanese ODA support that I studied, including the Bili-Bili Dam in Indonesia (Sato 2021). Nonetheless, my research indicated that farmers located downstream of these facilities experienced a notable increase in productivity and expanded economic opportunities due to the dams.

Table 5.1
Perspectives on irrigation projects

Perspectives	Examples of Interventions
Productive Work	
Transportation and transport	Multipurpose dam construction by the state
Agricultural production	Water regulation
Energy production	Power generation
Preventive Work	
Flood prevention	Mobilizing labor for regular dredging
Dredging to prevent erosion	Designing water flow to prevent malaria
Hygiene initiatives (preventing malaria and water pollution)	Communication and coordination with public health departments
Preparatory Work	
Research and analysis	Mapping water resources to identify areas that would benefit from irrigation
Civil engineering and establishment of a personnel management system	Establishing water user associations and management systems
Coordination between ministries	Communication with health and interior ministries
Compensation for forcibly resettled residents	Securing relocation sites and negotiating housing and compensation terms
Governance Work	
Conflict resolution	Establishing water user associations to mediate disputes
Ensuring a stable food supply	Balancing the interests of upstream and downstream water users, stabilizing urban food prices
Preparing for and managing disasters	Creating evacuation systems, revising settlement plans, and restructuring infrastructure

Source: By the author.

"Preventive" work encompasses interventions aimed at preventing environmental degradation from flooding and similar issues. The state is highly motivated to engage in this type of work, as it aims to ensure a stable food supply, prevent unrest, and keep citizens safe.

"Preparatory" work pertains to the administrative groundwork necessary for constructing large-scale infrastructure. This includes tasks such as research and analysis for financing and land expropriation. For these interventions, local individuals are expected to cooperate with the state. To secure funding, for example, the state may need to delegate the collection process to local communities, obtain loans from banks, and engage in other financial arrangements (Sampath 1992).

Finally, there is "governance" work. This refers not to the activities that produce the direct output of irrigation, such as water distribution, but to interventions that address the social consequences of irrigation, such as mediating conflicts over water. Traditional self-governing organizations within communities are often ill-equipped to resolve conflicts between upstream and downstream users. When such conflicts arise, locals must rely on government officials to mediate.

Discussions about maintenance typically focus on how to keep infrastructure operational in a sustainable manner. However, the core issue is how this infrastructure affects people's vital interests through the distribution of benefits and burdens. Failure to strike the right balance among communities can disrupt maintenance efforts, ultimately preventing the infrastructure from serving its intended purpose.

The larger the irrigation project, the more challenging it becomes to balance stakeholder interests. In his thorough historical analysis of Thailand's irrigation infrastructure modernization, anthropologist Hanten Brummelhuis explains how conflicts manifest in various ways; in one example, "navigation required a permanent highwater level, but rice growers preferred a low water level during harvest time" (Brummelhuis 2005, 154–155). As increasingly smaller canals branch off from the main arteries of irrigation systems, the effects of upstream conflicts cascade downward. This is why the jurisdiction of government-created irrigation offices often extends to downstream areas: Irrigation infrastructure inherently carries a host of social implications.

In recent years, many government-backed water user associations (WUAs) have emerged in rural areas across Southeast Asia. This sudden proliferation of WUAs fills a gap in organizations capable of handling the long-term maintenance of large facilities. Indonesia serves as a notable example. With few exceptions, such as Bali, WUAs in Indonesia are not grassroots initiatives; instead, they were established under government direction to support the construction of large-scale irrigation infrastructure. In regions of Indonesia that have operated irrigation systems for years, water is distributed through a fixed-turn system, allowing farmers to irrigate their crops on a rotating schedule managed by traditional water managers, known as *Mandoro Jene* (Watergate Keepers) who operate based on community trust (Oguni 2016). Such fixed-turn systems are often integrated into government-backed WUAs upon their establishment.

In Thailand, most of the "traditional" water user organizations were established only after extensive government-led irrigation infrastructure was built upstream, making water available to downstream communities (Ishii 1978, 21). These water user associations (WUAs) were formed under the direction of the Royal Irrigation Department, a government agency—evidence of the state's eagerness to gain the cooperation of local residents (Tomosugi 1976, 141). The primary purpose of these WUAs has been to secure low-interest loans, allowing farmers to avoid the higher rates set by commercial banks. By 2011, there were over twelve thousand WUAs in Thailand, with varying levels of activity among them (Ricks 2015).

Once governments establish new lines of contact with rural populations through WUAs, they gain a new means of exerting power: the ability to grant or withhold crucial funding. This dynamic exemplifies how resource management can create networks of dependence, entangling both societies and states in complex relationships of power.

5.4.2 Power Runs Deep: The Case of Indonesia

It should be clear by now that the relationship between the state and society is never fixed, even in the management of hard infrastructure. Change is driven by local capabilities, demographic shifts, and evolving sources of livelihood, but the level of state involvement is also a crucial factor. State power penetrates local communities in four stages: initial intervention, continuous involvement, naturalization of involvement, and finally, exit or extrication. Table 5.2 illustrates how this process unfolded in Indonesia.

Initially, Indonesian communities managed water use through a water manager–led fixed-turn system, though water-abundant areas such as Bali and certain higher-elevation regions employed different approaches. During Dutch colonial rule, irrigation infrastructure

Table 5.2
Drivers and bottlenecks for state involvement in irrigation

Stage of state involvement	Drivers	Bottlenecks
Initial intervention	Construction of infrastructure; increase in production	Resistance from indigenous leaders and so on
Continuous involvement	Infrastructure maintenance	Cooperation from local communities
Naturalization of involvement	Support from local community organizations (government-led water user associations and so on)	The need for the government to secure funding
Exit/extrication	Decline of the agricultural sector; growth of private companies; possible development of civil society	Monopolization of authority and responsibilities by the state

Source: By the author.

was developed along similar lines. It was not until the 1970s that postcolonial Indonesia began to develop extensive irrigation infrastructure, coinciding with the farmers' transition to high-yield rice varieties (Lorenzen and Lorenzen 2008). Also known as the "Green Revolution," centralized irrigation systems, together with intensive use of chemicals and high-yielding varieties to boost agricultural output beginning in the 1960s, ignored traditional small-scale irrigation. Therefore, it not only disrupted water schedules but also invited uncontrollable pests and diseases, resulting in crop failure (Lansing 1991).

Operating and maintaining the new centralized irrigation system imposed significant burdens on the state as well. The state sought to alleviate these pressures by forming WUAs across Indonesia from which it could collect water fees. By 1992, according to the FAO, four hundred irrigation systems in Indonesia had been transitioned from traditional setups to state-supported WUAs (FAO 1993).

In the second half of the 1990s, the Indonesian government transitioned from continuous involvement to a more naturalized role. This shift often marks a point where WUAs become more widespread but also drifted away from their stated purpose of representing all water users. For instance, a survey conducted in the Kulon Progo Regency of the Special Region of Yogyakarta, Indonesia, from 2003 to 2007 found that state-led WUAs facilitated collusion between wealthy farmers and officials, with the latter openly favoring affluent farmers in water allocation (Suhardiman 2018, 48). Despite the role WUAs play in upholding traditional power structures, the government frequently routed all water-related interactions with citizens through these associations (Hirayama 2016).

In summary, large organizations similar to WUAs were absent in most areas of Indonesia in the past; water allocation was managed by water managers. There were few exceptions, such as Bali's traditional *subak* water distribution system. A clear conflict exists between the government, which aims to conserve resources and reduce labor costs by delegating maintenance to local institutions, and farmers, who want government support in maintaining irrigation infrastructure. This conflict manifests not only in Indonesia but also in other Southeast Asian countries.

In Indonesia, the government has yet to extricate itself from the maintenance of irrigation works despite increasing delegation to WUAs. It seems unlikely that the Department of Irrigation will relinquish the authority and responsibilities it has assumed, given that most irrigable land has already been developed and maintenance is now as crucial as constructing new facilities. Furthermore, agricultural production alone will not dictate future irrigation-related needs; the prevention of floods and landslides has become an increasingly pressing issue. Climate change also exacerbates pest problems through warmer temperatures that significantly alter seasonal activities of insects and their population dynamics, damaging agricultural crops throughout Southeast Asia.

The challenges of water resource management are universal. Water flows downhill everywhere, compelling all countries to address irrigation needs to prevent floods and support agriculture. The methods that countries and regions employ to manage water reflect their social systems in fascinating ways. As demonstrated, societies respond to similar geographical and physical challenges in diverse manners.

5.5 How the State's Power Infiltrates Rural Areas

5.5.1 Infrastructure Inviting Inversion

One peculiar feature of the environmental state, which exercises its power through infrastructure, is its tendency to expand without a deliberate pursuit of increasing its realm of influence. While developmental states are often thought to be guided by intentional planning, the environmental state that emerges is largely an unintended consequence. Irrigation serves as an excellent example.

In chapter 3, I highlighted how the Japanese and Thai states retain and consolidate their power over forest and mineral resources in rural areas. The state's management of these resources is influenced by the ease of their commercialization. For example, states may invite foreign companies focused on export development. Water resources, the topic of this chapter, present a different scenario.

Water is an inherently decentralized resource that states find challenging to manage directly from the center. It is essential resource for both the wealthy and the poor, yet no one can fully monopolize it. However, when dams and other large irrigation facilities allow government officials to control the amount of water released into arterial and primary canals, water-dependent farmers are incentivized to seek preferential treatment—sometimes resorting to bribing the officials. Moreover, allowing arbitrary actions by individuals upstream, such as altering the volume or timing of water discharges, can have cascading effects downstream.

Figure 5.3 shows a typical irrigation structure serving as floodgates controlled by the local community. If we temporarily overlook the risks mentioned, the ability of farmers to cultivate and harvest crops even during the dry season offers significant economic advantages. After the completion of the Bili-Bili Dam in southern Sulawesi, one of Indonesia's main islands, annual rice production in the dam's basin more than doubled, increasing farmers' monthly incomes by a factor of 2.4 (Japan International Cooperation Agency 2016). It is easy to understand why farmers living downstream of dams tend to welcome state-led water development projects unless these initiatives involve forced relocations or similar disruptions. Once irrigation infrastructure is established, farmers often feel compelled to assist with its maintenance, drawing them further into the state's sphere of influence.

It is important to note that states still encounter various obstacles in constructing substantial irrigation works. Residents may entirely oppose forced relocations or may fight for higher compensation than what the government offers. Farmers who rely on small ponds

Figure 5.3
Flood gates between secondary and tertiary canals in southern Sulawesi. *Source*: Photograph by the author (September 2018).

disconnected from state-constructed waterways may be unwilling to maintain infrastructure they do not use. Thus, there are numerous opportunities for these projects to provoke backlash.

When does the management of water resources become a cause of inversion? Typical triggers include governments forcibly replacing community-sustained irrigation systems or reneging on their obligation to maintain infrastructure. Governments often justify their attempts to shift maintenance burdens onto locals with slogan-like expressions such as "resident participation" and "decentralization." Condemnation of "inefficient" and "premodern" practices often accompanies these government actions, which can lead to forced migration (Pudjiastuti and Wong 2023).

States can also invite inversion in other ways. For example, government-led projects that alter the natural flow of waterways can decrease the amount of water in natural reservoirs located near the outer ends of irrigation canals. Many of these ponds are too small to appear on maps, and governments rarely take them seriously in policymaking. However, they can be critical to local communities—for example, for watering livestock (van der Meer 1988). The state can also subsidize the use of pesticides to match production capacity in locations where more water becomes available by irrigation. As we briefly saw in the failed case of the Mega Rice Project in Indonesia, such intervention tends to have a shorter life not only because of the environmental disruption it causes but also because such destruction alienates local people from the region.

When states violently disrupt the autonomy of a local community, inversion is triggered, causing locals to lose their incentive to cooperate with the government. However, state power can also be exerted in less overtly draconian ways. Sociologist Michael Mann defines "despotic power" as the institutionalized and routinized force that state elites can wield without negotiation with civil society (Mann 1984, 188). Above all, Mann highlights "infrastructural power" as "the capacity to actually penetrate society and to implement logistically political decisions" (Mann 1984, 189). In other words, this is the power to influence civil society through the social infrastructure that forms the foundation of communication and transportation. Once embedded in a system, infrastructural power is difficult to overcome. The need to maintain and manage these facilities contributes to continuous contact between state power and citizens. In this context, neglecting one's duty to maintain infrastructure is not so much a resistance against the state as it is a violation of communal responsibilities. By integrating itself into the power relations connecting local communities, the state's influence can take deeper root in a region.

If locals resist state intentions to extend waterways and boost production, wouldn't this ultimately harm the interests of the farmers themselves? If so, isn't there an incentive for the farmers to prevent inversion from occurring? The answer to this question brings us back to the structural characteristics of irrigation. If upstream farmers abandon the facility, there is little that downstream farmers can do to access the water. The unique features of the irrigation structure make inversion almost an inevitable phenomenon. This is why the initial design of interdependency is crucial for the sustainable use of water resources.

In summary, the state creates a range of social infrastructure to ensure a stable supply of the resources it needs for economic development. To maintain this infrastructure, the state interacts and negotiates with local societies. It is challenging for the state to unilaterally compel people to maintain infrastructure; instead, it tries to convince locals to cooperate willingly so that the cost of conflicts and imposition will be reduced. These same locals, on the other hand, must rely on the state's power, even as they seek to resist it. In other words, inversion is triggered by a countermovement of local people aiming to restore the interdependency disrupted by state power. The proactive resistance sometimes displayed by the people is an effort to rebalance this unequal interdependency. For example, people in peripheral areas greatly benefit from new public health centers established to address the soaring cases of malaria that result from the construction of irrigation ponds where mosquitoes thrive. To ensure national security, states prioritize managing infectious diseases and serious health issues. In this context, the needs of the government align with those of the residents, creating an opportunity for a new kind of interdependency formed through ongoing negotiation between the state and its citizens.

However, this relationship is never equal, as the government holds more power than the people in defining the direction of infrastructure development. For instance, through its ability to design arterial canals that control the volume and timing of water discharges, the government establishes itself as a dominant presence in the lives of farmers. Thus, there is an imbalance between the significant potential for interdependence and a latent tendency toward dependency. As we shall follow up in chapter 6, the local people's ability and eagerness to unite and strengthen mutual support, instead of relying on the state, is an expression of "resistance" to revitalize interdependency within the community rather than with the state.

5.5.2 Forcing Cooperation to Preempt Inversion

This chapter has explored the processes by which the state engages with local communities. As the state intensifies its resource usage, it strengthens its control and generates a host of new tasks that require constant attention. Analyzing how these tasks are divided between the state and local communities may also reveal how the shock of inversion can be mitigated by carefully allocating authority in areas where responsibility does not reside with either the government or the people. At the very least, a thorough examination of the interdependence between state and society may allow us to move beyond a simplistic state-versus-local-community framework.

Is there a way for governments to prepare for the eventuality that their policies will lead to conflicts among farmers? Several options exist. For example, governments can ensure that water is not used until all downstream irrigation facilities are fully operational. Alternatively, they can grant preferential water use rights to farmers downstream. In southern Sulawesi, where I conducted research, authorities implemented a water distribution system that prioritized farmers at the downstream end of the irrigation canals. However, this system was only established after experiments demonstrated that there was sufficient water for all farmers and after the government facilitated an agreement on water distribution between upstream and downstream villages. When the state resolves conflicts between villages by equitably allocating benefits, it increases interdependency among local people and incentivizes mutual aid. In other words, it creates an "intermediate" sphere between the government and individual residents, serving as a buffer against unexpected conflicts.

Another tactic available to governments is to impose a "duty of maintenance" that encourages citizens to collaborate with one another. This approach can manipulate states to unite groups of farmers with differing interests. Looking at figure 5.4, a modified version

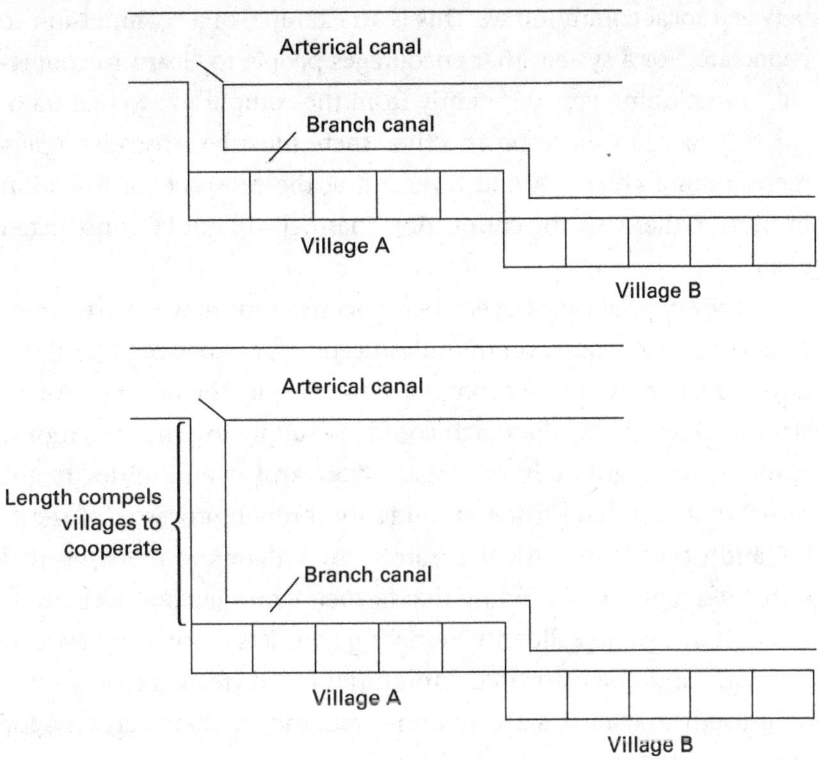

Figure 5.4
Compulsion to maintain. *Source*: Modified by the author based on Keohane and Ostrom 1995, 142.

of figure 5.1, consider what might occur if the design of the branch canal connecting the arterial canal to the field channel is lengthened, as depicted in the image at the bottom. This longer canal will become clogged more easily, meaning that unless farmers from both villages work together to keep it free of silt and debris, neither upstream nor downstream farmers will obtain the water they need.

A longer canal is less efficient for water users due to the increased risk of leaks and evaporation, as well as higher maintenance costs. Moreover, these increased maintenance costs will create a situation in which upstream village A becomes dependent on the financial and labor contributions of downstream village B. In other words, this government-imposed solution drastically alters the relationships between local communities. This is an example of a "compulsion to cooperate," or a system that encourages people to "learn to cooperate," functioning very differently from the compulsion to maintain. For this mechanism to be effective, there must be a broader agreement among villages A and B regarding the prospects of irrigation projects. Otherwise, the connecting channel will not be constructed at all.

Is it appropriate for governments to redesign new waterways to ensure that upstream communities become so dependent on those downstream that they are compelled to consider the needs of downstream villages? This approach could also apply to other situations. A more thorough study of already close-knit communities might reveal that they have fostered solidarity through processes of resisting and negotiating with the state or by welcoming much-needed state interventions. Assuming that farmers can wield political power through unity and solidarity, resolving conflicts of interest between upstream and downstream communities could create a strong force of resistance against state tyranny, enhancing their capacity for cooperation.

This chapter described how the need to maintain large-scale irrigation works can trigger inversion. Southeast Asia is widely viewed as the region most likely to be crushed by climate change, where people rely heavily on agriculture and fisheries. The compulsion to maintain the irrigation infrastructure to combat such challenges compels governments and local communities to forge new relationships. This water-mediated tightening of bonds between the state and the people will persist as climate change leads to increasingly frequent extreme weather events. Adopting the "maintenance" perspective will help us understand not only the relationship between nature and society but also how states will engage with local communities in the future through "learning to cooperate." Such an understanding forms the foundational basis for precautionary policies against inversions.

6
Solutions: The Promise and Challenge of Intermediary Groups

We have observed various patterns of inversion, where the state's objective of conserving the environment and pursuing sustainability is often undermined. Where can we find a way out? Even if complete prevention is inconceivable, is there a way to minimize harm so that local communities can work effectively with the state? By invoking classical insights from social analysts such as Émile Durkheim and Alexis de Tocqueville, this chapter reevaluates the role of intermediary groups as a potential solution to inversion.

6.1 Community and "Decentralization"

Intermediary groups provide platforms for individuals to form sustainable interdependencies; they facilitate discussions, coordinate actions, and resist external domination when necessary. Inversion is most likely to occur in situations where isolated individuals are connected individually to the state, leading to increased dependence. Without support from local groups, people are vulnerable to manipulation by the state. Typically exemplified by labor unions, cooperatives, voluntary organizations, and religious associations, intermediary groups occupy a crucial position between the state and the individual. The term "civil society" has a similar connotation, but I prefer to use "intermediary groups" because it explicitly focuses on the interface between individuals and state power. More specifically, the generic label of civil society obfuscates the fact that *each* intermediary group has its own independent mission and services it provides to society. The concept of intermediary groups thus allows us to focus on their distributional impacts (e.g., who is included in membership and who is not), giving us a clearer view of their roles in preventing inversion.

The strength of these groups can be critical in avoiding inversion. Whether the issue is sustainability, environmental concerns, poverty, or inequality, conventional thinking often revolves around the roles of the state and the market, analyzing the balance between the two. This perspective is grounded in the idea of social contract emphasizing the autonomous agency of each party involved. Discussions surrounding social contracts generally focus on the provision of benefits, such as income, education, and health services, raising the question of who should be responsible for these provisions. Social contracts are viewed as partnerships between individuals, businesses, and civil society organizations aimed at collective benefits (Shafik 2021, 10). As we shall see, intermediary groups can help safeguard against the risks faced by individual members of society.

In this chapter, I discuss the functions of intermediary organizations that challenge the social contract assumption of reciprocity among autonomous agents. Rather than focusing on what intermediary groups can provide, I emphasize their ability to shield communities from external threats, such as the encroachment of centralized states on local autonomy and resource exploitation. Guided by the historical struggle of local communities against powerful external forces, I concentrate not on what the state can offer or assign to these groups but on the importance of safeguarding them. While social contract theories assume that actors are independent, autonomous, rational, and reciprocal decision-makers engaged in the exchange of goods or services, real-life relationships of care—such as the bond between a parent and a child—are intrinsically asymmetrical (Noddings 1984). Intermediary groups can be viewed as an extension of such care, expanding beyond families to encompass the broader nexus of political economy (Gopnik 2023).

Natural resources such as water and forests are foundational to the livelihoods of people and communities worldwide. The history of control over these resources is deeply embedded in the struggle against centralization, domination, and exploitation by those in power, often ignoring how existing communities have maintained, conserved, and benefited from them (Haber and Menald 2011). The use of natural resources—such as extracting timber from forests and water from watersheds—requires substantial time and labor for cultivation and maintenance, which creates an interesting affinity with the concept of "care" in the asymmetrical relationship between those who nurture and those who consume. Like parents raising their children without expectation of returns, foresters plant trees that they often will not harvest.

While there has been significant progress in our understanding of how communal resource management operates on the ground (Ostrom 1990; Saunders 2014), there is still a need for a broader investigation into the nature of intermediary groups that carry out such governing practices. Conflicts arise precisely when state interests emerge in appropriating communal resources—whether for economic or political reasons—that have long been outside the state's purview. Throughout the long history of tension between those who govern and those who are governed, coercive techniques of government have evolved into more subtle processes of governance, often accompanied by language such as "citizen participation" and "transitional justice." As illustrated in chapter 2, these processes can either empower or disempower local knowledge and perceptions.

In developing countries, "empowering" initiatives are typically led by international aid agencies. While it is understandable that the international community seeks to decentralize power in pursuit of democratic governance, the motivations of nation-states in supporting decentralization are less clear, particularly since such a move might disempower the state. This hurdle to decentralization has long been recognized in development studies, which explains why proposals for land reform seldom materialize when power is primarily held by land-owning interests (Hirschman 1967, 154). A frequently overlooked aspect of the debate is the possibility that communities themselves may become obstacles to empowering decentralization. Paulo Freire articulates this point eloquently: "In 'community development' projects, the more a region is subdivided into 'local communities' without a thorough examination of these communities as both independent entities and parts of a larger whole (such as the area or region)—which itself is part of an even larger totality (the nation, as part of the continental totality)—the greater the intensification of alienation. As people become more alienated, it becomes easier to divide and keep them divided" (Freire 1970, 138).

Freire observed that overemphasizing "community" can prevent people from uniting and forming networks of communities. "Community development," he argues, has the paradoxical potential to weaken communities in the long run.

Why, then, do we sometimes see people cooperating with the state and committing themselves to conservation and sustainability, even when the benefits are not directly perceivable? Drawing on Michel Foucault's concept of governmentality, political scientist Arun Agrawal suggests that this shift is not simply a spontaneous change in people's attitudes (Agrawal 2005). Instead, it is facilitated by the transfer of authority to the local level, where many individuals become willing participants in conservation efforts. "Decentralization" has successfully reduced the cost of forest management, transforming villagers into collaborators. According to Agrawal, this specific form of decentralization has mitigated uncertainty regarding the relationship between forests and people, making both of them easier subjects govern from the political center (Agrawal 2005).[1] Whether the relationship between the state and the populace becomes inverted depends significantly on the autonomy of intermediary groups—comprised of individuals—relative to larger entities such as the state.

6.2 Groups Mediating Environmental Conflict

A tragic example of the function of intermediary groups can be found during the World War II in Japan. The infamous *akagami* (draft notices in red paper) were sent to summon the youth of Japan to the front lines. Although local village offices delivered the red papers to each household, it was a group called the Neighborhood Insurance Organization (*rinpo soshiki*) that made the system inescapable and binding. This organization held collective responsibility for the communal activities of the neighborhood. The five-person *rinpo soshiki* system created in the Edo period was originally intended to assist the poor, including the aged and childless, or the young and fatherless, rather than to serve the restrictive function described here (Tachibanaki 2017, 70). However, the original purpose of this organization was subverted. Through these neighborhood associations, the military relied on residents as "eyes" to crack down on draft dodgers and ensure broader cooperation in the war effort. These local groups linked individuals to the nation, which demanded troops to support its cause. The co-optation of such local groups has the potential to intensify strife and violence when they become entangled with the central state.

In contrast to these coercive mobilizations of intermediary groups driven by national interests, a variety of bottom-up organizations provide mutual aid and support. Many cooperatives and religious organizations operate in this way. While competition and division of labor—a defining characteristic of a market-based economy—can be seen as social mechanisms that combine individuals' strengths, intermediary groups not only harness these strengths for collective purposes but also accommodate weaknesses and offer mutual care. Local temples and churches, for example, are traditional intermediary groups that have fulfilled this function.

Émile Durkheim was one of the first scholars to discuss the significance of such groups situated between the state and the individual.[2] He referred to them as *groups secondaires* (secondary groups), including families, churches, political organizations, professional associations, and more (Durkheim 1974). Throughout much of human history, when states were not as dominant and influential as they are today, these traditional intermediary groups were indispensable in meeting the livelihood needs of both urban and rural populations.[3]

Durkheim focused on intermediary groups because he believed they could serve as a nexus, uniting the various professional groups that had become disjointed within society. He characterized the delicate relationship between intermediary groups and the state as follows: "We have seen that these secondary groups are essential if the State is not to oppress the individual; they are also necessary if the State is to be sufficiently free of the individual. Indeed, we can imagine this as suiting both sides, for both have an interest in ensuring that these two forces do not remain in immediate contact, even though they must be linked to one another" (Durkheim 1957, 134).

By the state being "free of the individual," he means that the state does not have to respond to the minute and varied local needs within the country it governs. By belonging to an intermediary group, the individual gains freedom in one aspect while becoming restricted in another. The expression of an individual's diverse needs is influenced by their experiences and discussions with fellow group members, shaping the balance between individual freedom and restriction.

As someone who grew up in the city, I was always amazed by the variety of voluntary groups in rural Japan. These groups include traditional community organizations such as firefighting volunteers, youth groups, women's associations, agricultural cooperatives, fishing cooperatives, and various local non-profit organizations (NPOs) dedicated to public interests. Some of these intermediary groups focus solely on improving the welfare of their members, while others pursue a broader public interest. What struck me most was that individuals in the countryside often belong to multiple groups simultaneously.

In this chapter, I limit intermediary groups to those that meet the following criteria:

1. They are positioned between the state and the individual, possessing the ability to voice concerns and negotiate with the state or governmental bodies, while also being oriented toward protecting the interests of their members.
2. They are essentially nonprofit and retain voluntary membership through camaraderie.
3. They have open membership not limited by social status or any hereditary attributes.

Focusing on sustainability in developing countries, I chose not to broaden my criteria to include private companies and related associations that are positioned between the state and individuals. While such organizations may function as intermediaries, their profit motivations distinguish them from the intermediaries I address concerning inversion. Although the precise definition of what constitutes an intermediary remains ambiguous, these criteria suggest their likelihood of committing to a particular place, in contrast to companies that tend to relocate and select members based on profit calculations. Commitment to a place is key to protecting local resources and individuals in the long term.

In short, the role of an intermediary group is to protect its members while providing public services such as safety, disaster prevention, environmental preservation, education, medical care, and town development according to local conditions. If powerful outsiders interfere with their activities, these organizations potentially possess the motivation and capacity to fight back. What qualifies such organizations as intermediaries is determined by their relationship with surrounding entities and by the interconnections among the individuals they represent. In other words, they establish a nested relationship with a larger organization, such as the state. Unlike the standardized systems that offer a bird's-eye view of their subjects and impose external (dis)incentives, these intermediary groups pursue their organizational objectives based on the voluntary will of their members and the contextual needs of each locale. More importantly, the vitality of intermediary groups lies in *doing* something rather than merely *being* there.[4]

In the introductory chapter, I briefly highlighted documentary film *The Warriors of Qiugang*, which depicts the intermediary roles of the local village, the NGO, and the media in the fight against a polluting factory in rural China (Yang and Lennon 2011). The film treats the village as a strong base from which to form social movements against the powerful factory backed by the local government. Collective action by the villagers demonstrates a strong commitment to their community. Conversely, the film also reveals how the structure of interdependence can invite *inaction*, allowing the situation to remain unchanged despite its severity from the perspective of the vulnerable. Most affected villagers are reluctant to raise concerns, fearing it might disrupt their community ties and peaceful village life. A shift occurs when an NGO arrives to support the villagers in making their claims. The mere existence of intermediaries does not guarantee their effectiveness as agents of change. Rather, they should be seen as potential catalysts through either open media or direct communications with state authorities and civil society organizations.

6.3 Why Intermediary Groups?

We have devoted space to defining and developing the concept of intermediary groups. This explanation is crucial for understanding sustainability because many natural resources, such as water, forests, pastures, and fisheries, are managed by intermediary groups as local commons, often based on communal customs and local rules (Ostrom 1990; Saunders 2014).[5]

Local commons can encompass a range of collective resources that resist monopolization, from forests and lakes to fishing grounds and pastures. Universally accessible commons can serve as a genuine lifeline for poor inhabitants in rural areas where private resources are scarce. In Thailand, for example, local commons fulfill a wide range of functions: They regulate the environment, help to conserve water and prevent floods, and serve as a source of local consumables such as herbs, mushrooms, game, fish, and firewood. Importantly, the commons are shared land reserves that can be utilized when private arable land in a region is insufficient (Shigetomi 1997).

Literature on communal resource management has identified key principles of sustainable resource use, such as clearly defined boundaries, graduated sanctions against rule violators, and low-cost means for dispute resolution (Ostrom 1990). These rules have primarily been discussed in the context of preventing the emergence of free riders within local communities. While it is challenging to create and maintain such local practices, the relationship with outsiders—especially external authorities—can be even more complicated, as states are often the ultimate guarantors of local rules.

Such conflicts can escalate when uncertain events are anticipated, such as flooding, droughts, and pollution. An environmental state could emerge in response to a social movement that collectively acknowledges the negative impacts of industrialization and development, as we saw in the Minamata case in chapter 2. German sociologist Ulrich Beck, a leading expert on the politics of risk perceptions, referred to this phenomenon as "colonizing the future" (Beck 1998, 11). Returning to the central theme of this book, environmental states are often the most eager to colonize the future through their chosen risk mitigation measures. Various forms of intermediary groups that emerged to debate and protest against polluting factories in Japan as a spin-off from Jun Ui's *Kogai Genron* evening class were prime examples of resisting such a narrowing of vision (chapter 2). While these groups may not have been powerful enough to completely stop the operation of such factories, their value should be appreciated if they managed to slow down the imposition of these unwanted facilities.[6] Such slowdowns often provide space and time, allowing marginalized groups to adapt and form their own strategies for survival.

"Preparing for the future," a favorite slogan of governments, may sound rational and prudent; however, it suggests that decisions about the future can and should be made by the government and experts, regardless of their specific orientations and problem-solving approaches. This poses a danger, as it allows those in power to obscure their actions behind a veil of "uncertainty," co-opting the concept of risk to favor their own interests under the guise of "preparation." The state's authority to prepare for the future can easily conflict with the efforts of local communities. In such cases, inversion is most likely to occur. For example, the confiscation of local commons into government-owned forests illustrates this issue. While the state may prioritize the protection of future biodiversity, local residents may rely on those same forests for fuelwood and other resources.

Although local communities may attempt to address future uncertainties in their own ways, they often lack the resources necessary for adequate preparation. This is where the government plays a crucial role; one of its functions is to anticipate and prepare for disruptive events such as resource depletion, wars, and natural disasters. Unlike local communities, the state employs a specific logic and methodologies that prioritize national security above other considerations. While this does not imply that the state and local communities possess entirely divergent views on preparing for the future, significant differences may exist in their perceptions of the problem and the proposed solutions.[7] These discrepancies can lead to outright conflict when the state exploits uncertainty to protect its own future interests, often at the expense of local communities.

The state's actions regarding local commons provide a compelling example. Throughout history and across the globe, local commons have taken on a variety of forms. How have these lands been integrated into state-controlled systems, and how has this affected their intended purposes? In developing nations, the focus on economic growth and industrialization tends to promote an emphasis on private property, exchange, and wealth creation. These objectives starkly contrast with the fundamental principles of local commons: community ownership, direct consumption, and communal sharing. Globally, people, goods, and money circulate and operate on an international scale. However, local commons serve as an in situ source of vital support for rural populations, enabling them to resist state pressures when necessary. They are the *local* resource.

What is required for these intermediary groups to act as a bulwark of democracy while maintaining an appropriate distance from the state, so they do not become entangled in centralization or co-opted as pawns of the state? To understand this, we need to examine the complex relationships that intermediary groups forge between the state and local communities, as well as within each group. These multilayered relationships influence individuals' capacity to withstand external shocks.

A contemporary example may illustrate this. During my fieldwork in the Mekong region of Southeast Asia, I encountered a robust local group called Thai Baan Research (*thai baan* means "villagers" in Thai). The villagers involved in the decade-long struggle that led the Thai government to temporarily decommission the Pak Mun Dam in northeastern Thailand pioneered the Thai Baan methodology in 2001. Recognizing that traditional research methods often overlook the intricate relationships between local livelihoods, culture, and river ecosystems, these affected villagers took the initiative to conduct their own research on the impacts of reopening the dam's gates on their lives and the Mun River ecosystem (Mahr 2021, 107–120). The significance of this group lies not only in the villagers conducting research for themselves but also in the skills they acquired to engage and negotiate with outsiders, which subsequently opened access to previously inaccessible external resources. While they do not represent a case of overcoming inversion, they provide important hints about the enabling conditions that can produce successful intermediaries.

Let us take a step back and place intermediaries in a historical context. One important function of the state has been to mediate conflicts among various regional groups within its territory. In Japan, well before the establishment of the modern state, it was customary during the Edo period (1603–1867) to seek intervention from a higher authority to resolve disputes over trans-regional issues. For example, the distribution of irrigation water—a matter of life and death for farmers—often led to conflicts between villages. The shogunate generally played the role of "objective mediator of conflicts of interest" to find a solution (Watanabe 2014, 214). A peasant born into a village had to live within its limits and rely on its resources for survival, seeking assistance from outside authorities such as lords only when trouble arose. However, as will be discussed later, these intermediary groups were dismantled one after another during the process of Westernization and modernization.

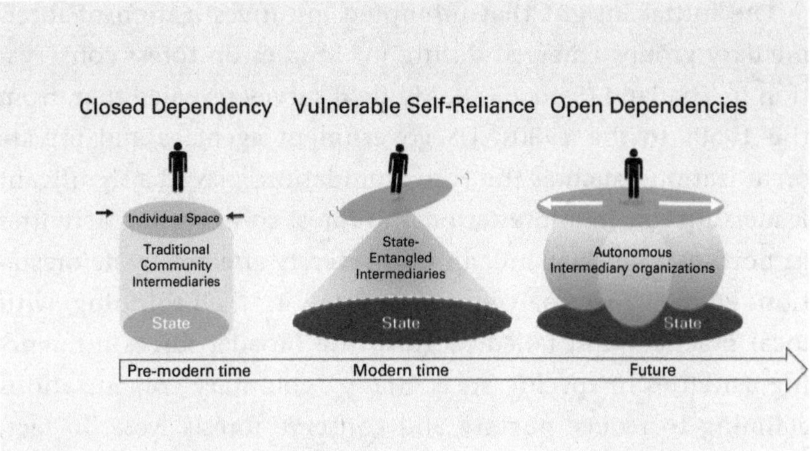

Figure 6.1
Traditional community intermediaries and state-entangled intermediaries. *Source*: Created by the author.

Figure 6.1 illustrates the transition of intermediary groups, moving from traditional premodern intermediaries characterized by "closed dependency" relations (left side of figure 6.1) to the modern era of state expansion, where state-created intermediaries are entangled with centralization (center). In the contemporary postmodern world, the social ideal of self-reliance—that is, the ability to "take care of yourself"—has become predominant, while the notion of dependency is almost universally demonized, especially in Western culture. As state influence expands and becomes increasingly intertwined with individual lives, no matter how "independent" people may seem, they often become mere cogs in a wheel, rendering their livelihoods paradoxically precarious. In this context, the "intermediary" is essentially a pawn of the state,[8] lacking the intrinsic incentives and sense of purpose necessary to improve the lives of its members.[9] As state power approaches complete domination, the only way to check it is to establish multiple intermediaries on which people can choose to depend (on the right side in the figure), a topic we will discuss later.

The initial insight that prompted my investigation of intermediary groups emerged during my studies on forest conservation in Thailand (Sato 2000). My field survey revealed that, from the 1960s to the 1980s, US government agencies and private organizations, such as the Ford Foundation, played a significant leadership role in afforestation and forest conservation activities in northeastern Thailand, an area severely affected by deforestation. Perhaps I was naively preoccupied, in the beginning, with local practices and failed to grasp the broader forces influencing activities in specific areas. Many "voluntary" organizations claiming to reduce poverty and conserve forests were, in fact, merely government "outposts" established by the CIA and the Thai Ministry of the Interior. Their true role—at least in Thailand during the 1960s—was to monitor and influence people to prevent them from joining the Communist Party and to draw them toward the government through development and aid (Puangthong 2021).[10] This leads us to the conclusion that while a spirit of volunteerism is an important requirement for intermediary groups, it is neither easy nor essential to determine its presence or absence. Similar to the difficulty we face in identifying "the true intent" of the state in controlling people through resource conservation efforts, intentions should not be central to our analytical approach.

If intermediaries can easily become tools of the state to infiltrate society, this undermines the bottom-up incentive for locals to conserve resources. Consequently, the remaining hope for resisting this inversion appears to lie in our simultaneous connections to multiple intermediary groups (right side of figure 6.1). Human organizations come and go, and voluntary groups without kinship or other obligatory bonds tend to have short lifespans. However, when individuals can rely on voluntary groups and adapt to their practices of camaraderie, they tend to create a more stable and resilient foundation against sudden external shocks. Joining multiple groups amplifies this protective effect. Before we explore the significance of renewed dependencies based on multiple layers of intermediaries, we should consider the potential limitations of these intermediary groups themselves.

6.4 Tyranny within the Intermediary Group

Localized conflicts in intermediary groups such as neighborhoods, companies, and schools can sometimes cause more harm than state violence. Deaths due to overwork in Japanese companies exemplify this phenomenon. A prominent commentator on the topic stated, "The reason why individual rights are not fully respected in Japanese society is not because state power is too strong, but rather because the power exercised *within* the intermediary community is too dominating" (Inoue 2001, 163, emphasis added). He argued that the fear of tyranny within the community can be more problematic than that posed by the state.

Durkheim raises the question of whether entrusting the state to resolve the violence of small groups scattered throughout the country "will not become more tyrannical" (Durkheim 1974, 98). He recognized the potential for violence and tyranny within intermediary groups. Durkheim was particularly concerned that these groups, which maintain some autonomy from the state, often reproduce the very feudal relationships they seek to avoid between the state and its members (Durkheim 1957, 61–62).

The state, positioned to govern intermediary groups from above, is expected to mitigate conflicts among and within these groups when their capacity to handle issues is exceeded. Durkheim believed that the tyranny of the state would be even more intolerable than that of intermediary groups because it would be more uniform and arbitrary, with little regard for the unique conditions of each locale. He stated, "The state, in today's larger societies, is so out of touch with special local interests that it cannot take into account the particular, regional, and other unique local conditions. So, when the state tries to regulate the various special interests, it can only succeed in doing so by strong-arming and twisting" (Durkheim 1974, 98).

While Durkheim acknowledged that intermediary groups can harbor the seeds of feudal domination, he argued that they are necessary to provide a counterbalance against the state, which can distort regional "special interests" and invites inversion. One promising approach to addressing these issues is to expand the range of intermediary groups that individuals can choose to rely on.

Even if intermediary groups proliferate, they may establish unequal relationships based on their political and financial power. To allow each group to maintain a certain degree of autonomy without one dominating another, the state must act as a coordinator and monitor the relationships among these intermediary groups. Ensuring the autonomy of the groups will be crucial in preventing this monitoring from becoming excessive. The relationship between the state and intermediary groups—and among the intermediary groups themselves—determines the likelihood of inversion. These groups function best as spontaneous organizations. Instead of providing additional support, what is needed is for the state (and the market) to create some space for these organizations to operate autonomously.

6.5 Three Functions of Intermediary Groups

As the state and the market increasingly expand their respective spheres, the conventional idea of "leaving it to the state" or "leaving it to the market" has become a standard institutional solution for addressing public issues. Consequently, the space available for intermediaries to exercise self-motivation appears to be narrowing. What role is left for intermediary groups?

Three functions distinguish intermediary groups from those under the state and the market, offering insights into how they can address the inversion problem. First, by fostering a sense of camaraderie and establishing a "code" among members, intermediary groups can proactively address conflicts before they escalate. When individuals, freed from their former communities, pursue private interests without restraint, an atmosphere of competition and individualism prevails, undermining community trust. Without internal regulations, "loopholes" may proliferate, leading to strife within the intermediary group.

To prevent such outcomes, certain forms of coordination can be grounded in local contexts. For instance, in popular tourist spots in Japan, tourist associations are often established to coordinate individual hotels and guest houses in the area. These associations help prevent "land liquidation," where the most successful hotel acquires the property of the least successful, resulting in consolidated land ownership (Kurumisawa 2004, 151–153). The association acts as a bulwark against highly capitalized companies or wealthy individuals who have no connection to the area.

Second, as nonprofit organizations, intermediary groups provide goods and services related to community health and safety that are not supplied by the government or the market. In doing so, they seek societal recognition for the legitimacy of their contributions. Their activities are diverse, encompassing physical and medical care, educational opportunities, and community safety initiatives. Community-rooted activities are expected to be more resilient to external shocks and upheavals. However, intermediary groups are not governments and are not necessarily accountable for the fair allocation of benefits and burdens among citizens because they operate according to their own norms regarding inclusion.

Third, intermediary groups can pursue global ideals such as environmental conservation and democracy while positioning themselves beyond state control. This allows participants and members to critically reflect on existing local perspectives. One example is the Regional Environment Center for Central and Eastern Europe (REC), an international NGO where I served as a board member for several years. Established in 1990 in Hungary, REC aims to enhance the environmental response capacity of Eastern European countries to the serious pollution that became apparent after the fall of the Berlin Wall in 1989. It was founded with support from Japan and the United States to collect and provide diverse information, conduct training and capacity building, and make policy recommendations. Unlike the forest conservation groups we observed in Thailand, this organization negotiates with each nation-state while also collaborating directly with individuals within those states to advance their environmental advocacy capacity.

REC's mission was not only to disclose environmental information and make policy recommendations but also to foster a healthy civil society in the Eastern European region through environmental initiatives and to promote democratization. In other words, rather than directly solving environmental problems, REC aimed to nurture organizations and individuals who could resolve issues independently according to the specific context of each country. The environment became a platform for engaging in critical discussions about existing modes of economic production and political decision-making. In the former Soviet region, where decisions were traditionally made top-down, newly formed civil society organizations acquired the knowledge and skills to make decisions bottom-up and worked to further disseminate that know-how (O'Brien 2010).[11]

International voluntary associations, such as the International Red Cross and Doctors Without Borders, can also be viewed as intermediary groups whose activities are guided by various global principles. These organizations operate "between individuals and nations" across borders, distinguishing them from entities like the United Nations, where norms and operations require collective agreement among member states. As global issues such as climate change give rise to conflicts, expectations for these autonomous intermediary groups, which exist outside the logic of any particular state, are increasing.

6.6 Resources as a Source of Power

What conditions enable intermediary groups to exert influence beyond their membership? To explore this, let us return to the national level and examine some examples. We are particularly interested in instances where workers in subordinate positions have utilized collective action to effectively *speak* to the ruling class.

In his book *Carbon Democracy*, political scientist Timothy Mitchell examined why coal miners' strikes were effective in improving working conditions in Europe from the late nineteenth century to the early twentieth century (Mitchell 2011). The workers engaged in the hazardous task of coal mining were positioned at the bottom of society. Despite their reliance on the mines for their livelihoods, they successfully organized and executed strikes to negotiate with corporate management. Mitchell analyzed this bargaining power as a result of the convergence of two conditions. First, the isolated work environment of the coal mines, which was difficult to monitor, allowed workers to exert control over their labor, providing them the freedom to launch strikes and engage in sabotage. Second, during this period, industrial society was critically dependent on the coal workers at the end of the supply chain (Mitchell 2011, 19–20). The geographical concentration of coal mining and distribution afforded the impoverished miners who gathered there an amplified political voice.

It was not merely the existence of legal procedures for protests and grievances that empowered this intermediary group to resist the state. While organizing as a group was beneficial, the critical leverage these workers held for sabotage was paramount. As a railman quoted by Mitchell remarked, "With two pennies' worth of a certain substance, used in the right way, we can make a locomotive unable to work" (Mitchell 2011, 21). Industries' dependency on energy, combined with the isolation of the workers and the geographical concentration of coal production, enabled workers to connect their demands with those of others, lending their arguments technical force that could not be easily dismissed.

The interdependency of actors enabled coal mine workers to forge multilayered relationships that helped them assert their demands to more powerful groups. The coal mining example should not be viewed as unique, as these workers occupy a "privileged position" by providing the essential resources that keep the entire industry operational. Similar instances of broad-based cooperation can be found in other sectors.

The history of intermediary institutions resonates in the present day. Attwood's study of the sugar industry in Maharashtra, India, for example, offers valuable insight into how the nature of resources impacts relationships among key producers within the agricultural sector in the context of developing countries (Attwood and Baviskar 1987). Cooperatives of sugarcane producers provided loans, technical training, seed and compost support, and other services to their members, regardless of their economic stature. Despite the presence of large growers and domineering leaders of the sugar factories in Maharashtra, these cooperatives facilitated broad cooperation across different classes involved in sugarcane production, from production to distribution, even among the more vulnerable segments of the population, like smallholders.

Attwood and Baviskar note that this cooperation across social classes was not rooted in traditional village culture but rather in a shared interest in the goods and activities necessary for continued production. Cooperation is crucial, given that sugarcane is time-sensitive and perishable; if not distributed promptly, its value can diminish to zero. The perishable nature of certain agricultural products effectively compelled stakeholders aiming to profit from sugar production to collaborate with both small and large farmers. However, in regions dominated by a few large landowners, there is little incentive for such cooperation. The cooperation that transcended differences in position was necessary for the system to function, as no single association was strong enough to overshadow the others.

Again, these examples are not exactly the cases where inversions have been overcome. Yet, they provide important hints about the significance of interdependencies that confer voice and power to the actions of intermediary groups. Transforming dependencies founded on vested interests and entrenched class structures is challenging. However, regardless of how rigid these structures may be, the state and managers of large corporations must rely on the power of those on the ground who execute the jobs that the power holders require. The main reason the intermediary groups in these two cases were able to influence the state and private owners was that those in power were critically dependent on the people supplying resources on the ground. The fields of climate and environment exemplify this, as everyone from business executives to farmers plays a vital role. When the network of dependencies is assembled correctly, even those lower in the hierarchy have room to negotiate, creating opportunities for maneuvers that were previously unattainable.

We observed that the function of intermediary groups is influenced not only by the size of the organization and the communal sense of belonging but also by their original purpose. In the cases mentioned, the timely production and distribution of agricultural products and the assurance of workers' jobs and security were key bottlenecks. Intermediary groups are thus guided by an autonomous sense of direction that enables them to approach state power when necessary and to resist state intervention when threatened. Many groups—even those established spontaneously from the bottom up—are effectively incorporated into the state. Conversely, some groups that were created under state direction may eventually take ownership and continue on their own initiative. Intermediary groups can change their nature; for instance, the aforementioned REC has gradually drifted away from its original mission of "democratization through the environment."[12]

Creating and nurturing intermediary groups that can autonomously maneuver while maintaining a safe distance from the state is challenging. The size of a group is also significant: If it is too small, it may lose strength and influence; if it becomes too large, members may lose their sense of belonging. The balance between the forces of belonging (which are stronger in smaller groups) and the forces of fragmentation (which increase with group size) has been a central dilemma for traditional intermediary groups, such as religious congregations (Dunbar 2022, 97). To revive intermediary groups in the modern global era—where individuals can simultaneously belong to multiple groups—it is essential to establish a structure that does not rely solely on state subsidies.

The nature of intermediaries dictates not only how resources are used locally but also how access to those resources is regulated in relation to state power. Herein lies a seed for "inversion." If inversion arises from a fundamental conflict of interest—typically between local communities and larger organizations such as private companies and state bureaucracies—to borrow the German concepts, we can identify this conflict as that between *Gemeinschaft* (community) and *Gesellschaft* (society). The former emphasizes a sense of belonging, while the latter places greater emphasis on function (Tönnies 1957). While community can solicit voluntary sacrifice from its members, sacrifice in society can create more conflicts because members do not necessarily feel a sense of belonging. This mechanism becomes intense when members have no option but to endure the sacrifices.

6.7 Preventing Isolation, Despite "Less Splendor"

Nation-states are often too large for individuals to express their will and effectively expect changes to be made. Alexis de Tocqueville, who visited the United States from France in the first half of the nineteenth century, was impressed by how local groups called "townships" were organized to overcome this issue (Tocqueville 1863). Tocqueville coined the term "secondary powers" to describe associations such as townships and focused on their ability to independently oppose federal policies when necessary.[13]

The vibrancy of these local groups impressed Tocqueville enough that he remarked, "Municipal institutions constitute the strength of free nations. Town meetings are to liberty what primary schools are to science; they bring it within the people's reach" (Tocqueville 1863, 76). What kind of society can prevent individual freedom from becoming entangled in state power while simultaneously providing access to state resources when local issues cannot be resolved? Tocqueville saw the potential for democracy in grassroots groups found in townships, viewing them as a connecting hub for individuals to unite and gain strength.

A similar observation was made in a completely different context in modern Japan. Masao Maruyama, a leading political scientist of postwar Japan, used the term "atomization" to describe the danger of isolation. In response to the question, "Why do the masses enthusiastically support dictatorships?" he provided insights that illuminate this concern: "The atomized individual is often apathetic toward public affairs, but this very apathy can abruptly transform into fanatic participation in politics. This occurs because individuals, seeking to escape loneliness and insecurity, may identify themselves wholly with authoritarian leadership or immerse themselves in the mystical 'whole' represented by concepts like national community and eternal racial culture" (Maruyama 1968, 496).

To ensure that the state remains in the hands of individuals, it is essential to avoid creating atomized individuals who lack horizontal connections. Establishing a bastion of fellowship in the mind can act as a safeguard against losing a sense of direction. The sense of direction can be nurtured through more communication and openness to critical discussions rather than through exclusive reliance on limited information. The purpose of the association, therefore, is not merely to unite and strengthen one's political voice; it also serves as a compass to orient the self toward an educated imagination about desirable societal goals.

As discussed in this chapter, intermediary groups can serve as useful tools for state power, but they can also act as spearheads of resistance and deterrence. Local communities often have valid reasons to resist, as the resources they depend on daily, such as forests, pastures, ponds, and other local assets, have been integral to their lives for generations. While sophisticated divisions of labor can deepen human dependence through competition for individual advantages, intermediary groups are crucial entities that can embrace human vulnerabilities and provide safety nets. Given their multifaceted potential, these groups should not be viewed simply as volunteer organizations or NGOs/NPOs; rather, they should be seen as bulwarks that mediate between the state and individuals, preserving the delicate balance of mutual dependency between local communities and external powers.

Climate and environmental issues will increasingly become significant sources of conflict both between states and within each state. As these conflicts escalate, the state will find more opportunities to intervene and take control of issues that would otherwise remain local in nature. Conflict-ridden resources are central to state interests; the state may be less concerned about the sustainability of these resources and more focused on controlling the individuals who may rebel against state authority.

Intermediary groups play a critical role in mitigating state-society relations, thereby decreasing the likelihood of conflicts that can lead to violence against each other and the environment. Tocqueville provided a poignant description of the behavior of individuals in a society where the interests of the community align with those of the individual. In a democratic state, he remarked,

> if there is less splendor than in the halls of an aristocracy, the contrast of misery will also be less frequent; the pleasures of enjoyment may be less excessive, but those of comfort will be more widespread; the sciences may be less perfectly cultivated, but ignorance will be less common; the impetuosity of feelings will be repressed, and the habits of the nation softened; there will be more vices and fewer crimes. Each individual will feel the necessity to unite with fellow citizens to protect their own weaknesses; recognizing that they must cooperate to receive assistance, they will perceive that their personal interests are intertwined with the interests of the community. (Tocqueville 1863, 10)

Thus, intermediary groups that mitigate against inversion exist within the framework of the state but are not subordinated to it; they are part of the market system but not dominated by it. Essentially, they are groups that serve their constituent peers on the principle of reciprocity, as Tocqueville states, and they both free and constrain individuals. In the face of the exercise of overwhelming state power, these groups can and should serve as a bulwark of individual freedom.

It has been a long time since people lived in closed dependencies without choices regarding their reliance on other groups. In the past, individuals were often confined to the group into which they were born. Today, with increased freedom to move and relocate, people have a wide range of options. However, whether they can exercise these choices depends on their sociopolitical status and the societal acceptance of such decisions.

The significance of human dependency in the analysis of the environment and sustainability lies in the fact that human reliance on nature is deeply influenced by our mutual dependencies within human society. Expanding the choices available through intermediary groups formed by voluntary associations and enriching their activities will strengthen individuals' resilience against external powers that might otherwise foster inversion. However, as long as these groups operate legitimately within a democratic state, we cannot adequately assess the social contribution of each. It is more important to provide options for individuals to join intermediary groups than to focus on evaluating the legitimacy of specific groups.

Tocqueville emphasized that while there may be "less splendor," freedom will be more accessible as these intermediary groups are empowered. Strengthening intermediary groups involves reorganizing human relationships so that personal interests align with those of the community. There is, therefore, a compelling reason to restore the virtue of interdependency, even though it has long been demonized in the era of ever-expanding markets and states founded on atomizing social contracts.

Conclusion: Renewing Interdependency

> Independence should not mean doing things without someone else's help. Rather, it means having a network of people who will come to assist you when you shout out "help!" Such a state is not independence in a literal sense; it is more about interdependence with others. As this network goes beyond a single layer, it becomes more secure.
>
> —Washida, Kiyokazu, *Sense of Direction in the Dense Fog* (translated by the author)

7.1 Sustainability for What?

The quote above resonates beyond human society to include our interdependency with nature. It is high time to rethink how human relations are interwoven into our relations with nature. Unsustainable, and thus unsuccessful, environmental interventions, which I refer to as "inversion," highlight the consequence of our one-sided perspective of the entire system. When we focus solely on improving our methods for addressing threats to or from the natural environment, we risk deepening existing disharmony—not only with nature but also among people. We know from history that when states use their power to restrict access to natural resources for marginalized communities, they often perpetuate inequality, as seen in chapter 3. Despite perceptions otherwise, environmental policies are inherently political, from the problem definition stage to their implementation. The concept of the environmental state allows us to encompass a broad spectrum of these consequences, rather than remain fixated on evaluating individual policies based solely on their effects on the physical aspects of nature.

Just as some individuals view economic development as the paramount goal, others prioritize environmental conservation over other socioeconomic issues. Certainly, global society has come to recognize the importance of environmental sustainability after experiencing numerous harsh side effects of economic development, such as pollution and climate change. However, the simultaneous pursuit of development and conservation tends to favor the former, especially in regions where people's basic economic needs have yet to be met. Changes in US economic policy initiated by the Trump administration add another pattern to this, where domestic job creation at the expense of environmental concerns can still be prioritized in a richer country like the United States.

But what if the exclusion of local communities, as discussed throughout this book, was intentional, and the narrative of environmental conservation was merely a façade for political motives? This scenario becomes plausible when the state envisions the outright removal of local populations from sensitive areas—such as national borders or strategically important locations—primarily for political reasons. Environmental justifications might lend a technical veneer to such otherwise aggressive interventions. However, detecting these intentions can be challenging for outsiders. More critically, as illustrated in chapter 1, the state is composed of various ministries and departments with often conflicting interests. There is rarely a singular intention guiding policy implementation. This lack of cohesion is a fundamental reason I have focused the discussion in this book on the local effects of environmental policies, which shape the incentives for local communities to engage with or resist the state.

In contrast to the visible and contested environmental dimensions of economic development, the human dimension of environmental policy has long been neglected and remained largely invisible for various reasons. First, because the concept of sustainability was initially advocated by the United Nations, the "country" has become the focal point of efforts (Porntney 2015, 32). Various indices and statistical data used to compare sustainability performance, such as carbon footprints, are typically aggregated at the national level. This reinforces the perception of the "country" as the basic unit of sustainability in global discussions.

Second, because the broader consensus among countries has sidelined discussions about history—specifically, how the Earth became unsustainable—issues related to the costs and benefits of industrialization have been overlooked. By focusing on future goals, the sustainability discourse obscures the complex history of who benefited from the unsustainable development that we now come to criticize. Third, because the energy-intensive economic activities of people are often viewed as detrimental to the environment, indicators of success in achieving sustainability have been dominated by environmental metrics, often at the expense of human indicators. The sustainability of the environment must go hand in hand with the sustainability of human society.

On one end of the spectrum, radical environmentalists may advocate for population control to preserve the climate; however, such a solution is inadequate in the long term. Even in a country like Japan, where a shrinking population means decreasing damage to forests, the abandonment of previously cultivated human-made forests is becoming a threat, leading to potential landslides and other "natural" disasters.

I argue that while climate action is essential, we must also secure the right to development for human society, particularly for those deemed "latecomers" in the Global South. Without this, long-term environmental sustainability will be unattainable. The intensifying climate and environmental crises, along with their significant risks to all societies, compel us to rethink our definition of "development." This redefinition hinges on understanding the essence of development that ultimately determines the spaces needed for conservation, the reasons behind it, and the associated costs.

Since World War II, development has primarily been understood as the increase in economic activities aimed at growing per capita income and GDP (Coyle 2015). States designed to promote such development through explicit regulatory and public investment policies are referred to as "developmental states," a characteristic feature particularly of East Asian nations (Johnson 1982; Bresser-Pereira 2019). Currently, the transition from developmental states to environmental states is occurring at an exceptional speed among less-developed countries, especially in Asia, where global CO_2 emissions are rising at the fastest rate (Global Carbon Project 2024). As discussed in the previous chapter, the absorptive capacity of locally based intermediary organizations is not keeping pace with the relentless push for "development" in these countries.

Economic development reduces the time and distance associated with various economic activities through new technologies and reshapes societal competition. Additionally, it often converts environmental measures into just another facet of development, as seen in the race to develop clean energy (Thurbon et al. 2023) and in efforts to build resilient infrastructure for climate change adaptation. In practice, conservation is frequently integrated into existing development strategies. Consequently, developing countries are not always sidelining conservation; instead, developmental and environmental policies often reflect two sides of the same coin.

The primacy of "development" is deeply ingrained in modern societies. Development, whether technical, institutional, or cultural, is frequently seen as a solution to all social ills and is rarely perceived as a *problem* in itself. When hastily implemented environmental policies fail to yield the desired results, governments of developing countries—often with support from foreign donors—introduce new "plans," while failed policies are quietly assimilated into the system. As long as this developmental state framework remains dominant, the vicious cycle that leads to policy inversion will persist. Environmental policies are often hindered by misalignment. Many countries continue to subsidize fossil fuels, which inhibits fair competition with renewable alternatives, all while grappling with the regulation of resulting CO_2 emissions (Black et al. 2023). Why should people take emissions reduction seriously if the state encourages increased pollution?

As long as we rely on economic development and capitalistic competition to solve our existing problems, we risk perpetuating the very structures that created these issues in the first place. Development continues to advance, with many policies proposed based on specific calculations that draw from and reinforce the development mindset. We must consider: Do we have the tools to implement this policy? Do we have the budget? Will it be politically advantageous? Like clockwork, these calculations produce measures that recreate and exacerbate rather than solve problems.

The overarching question, therefore, concerns how we can untangle ourselves from this ever-expanding web of development that too often leads to inversion. The question of sustainability ultimately boils down to our development aspirations, which define what needs to be sustained. As we begin to understand the problem of inversion, I suggest that we create and implement policies explicitly designed to avoid causing more problems. In other words, we need a shift in mindset from solving problems to preventing them from occurring. I emphasize this because policy "problems" are defined by policymakers, not by the people who bear the consequences on the ground. Worse still, people are impacted not only by climate change but also by the "solutions" to climate change, such as restrictions on land use and resettlements.

If the developmental state continuously searches for what the country is "missing" and attempts to fill any identified gaps with more development, we must reconsider development itself *before* addressing the environment. The approach I propose is rooted in this shift in perspective and involves reflecting on the past while making the most of what we already have. It is certainly tempting to focus on deficiencies and shortages; however, this human tendency is precisely the blind spot exploited by external forces that come bearing "solutions" (Pritchett and Woolcock 2004). Examining how inversion complicates the solution brings us back to how problems initially emerge. Yet solutions within the scope of environmental policy are inherently constrained. So, what are our options?

7.2 Preventing Inversion

The emergence of environmental states is inevitable due to the rise of climate and other environmental threats that require a public response. These states should prevent the inversion created by their own policies. One option is to treat individuals who live in or near affected areas as primary stakeholders, and empower them to take the initiative in shaping policies. In other words, development must become decentralized to accommodate localized solutions. When projects are conducted on a local scale, regions are better positioned to mitigate against emerging inversions. For instance, projects could provide electricity to rural areas in developing countries by installing solar technology rather than by constructing large dams. Here, development is scaled down to be manageable for local communities. This presents a significant challenge in many authoritarian countries that are becoming dominant in Asia, but it is not impossible.

A second way to fend off inversion is to break down bureaucratic silos and integrate conservation into development policies. For emerging nations, this involves more than just improving the management of environmental projects. As highlighted in chapter 1, the administrative bodies responsible for development and environmental initiatives are often at odds. Environmental agencies regulate other government bodies but frequently lack the power to enforce their authority. The question then becomes how to effectively incorporate conservation mechanisms into existing administrative bodies responsible for development. The increasing frequency of natural disasters may create opportunities for bureaucratic shifts toward establishing a powerful agency responsible for sustainability that will not be dominated by development-oriented agencies.

A third approach to mitigating inversion is to enable intermediary organizations to respond to and adapt to changes brought about by societal, technological, or climate issues while avoiding the frustrations caused by solutionist policies. Inversion occurs when local communities lack the resilience to cope with external changes or shocks. While intermediary organizations may not possess the power to transform existing structures, they can slow down and serve as critical buffers against the external shocks created by reckless policies. There may be little that outsiders can do to build the intermediaries necessary for this approach to succeed. However, when such groups do emerge, they should be preserved and supported, rather than interfered with or controlled by the state. This is perhaps the best advice I can give to state policymakers. Environmental policies should start from what people already know and do. After all, the long-term goal of sustainability should not be in conflict with the interests of the people, wherever they live. The only question is whether such people are included in the purview of the state.

There is often a significant misalignment between the issues that states publicly support and fund, and the hidden consequences of these government initiatives. If this misalignment persists, the voices of the most vulnerable will remain unheard, and policymakers will fail to learn from those most affected by environmental policies. This study began by exploring the gap between the declared objectives of policies and their actual consequences. Environmental states will only learn and self-correct through external pressure resulting from the need for transparency and information disclosure. Conversely, when environmental states spiral out of control, the cause is often opaque policy implementation processes that create an information vacuum. If citizens never learn about activities on lands directly controlled by the state, they cannot criticize these actions or propose solutions. A more serious issue is that even when information is available, it often goes unexamined in public discourse. Journalists and the free press have an essential role in this regard.

This book has examined the ground-level sociopolitical consequences of the environmental policies prevalent under the current system. It argues that a mechanism is needed to reconnect on-the-ground events to the policymaking process, ultimately improving the quality of our natural environment while reducing instances of inversion. Our main challenges are accepting uncertainty as a given and maintaining robust processes for applying pressure to correct inevitable mistakes. We must foster the capacity for individuals to generate and wield pressure appropriately.

7.3 Environmental States as a Theory

A wide array of stakeholders with diverse interests is involved in the search for solutions to environmental problems. This has inspired a global movement to unite these actors behind a common goal, leading to the emergence of "environmental governance" in the 1990s (Lamont 1996). The premise of this concept is that the environment should be managed according to democratic principles developed through interactions among governments, companies, and civil society. Generally, scholars of environmental governance critique the conventional monopoly of the state over environmental management and seek ways for nongovernmental actors to gain meaningful power, fostering strong networks that support conservation. Essentially, the focus is on enhancing the resources available for effective environmental management (Bodin, 2017).

Theories of environmental governance typically focus on the structure of governance systems rather than the effects of strict conservation on local communities (Bodin 2017; Partelow et al. 2020). However, the impact of environmental problems is not uniform; similarly, the effects of "solutions" are not felt equally across communities. Take natural disasters as an example: While they may seem to affect everyone in a location equally, individuals' ability to cope is largely influenced by factors such as the quality of their housing and their financial resources. Subsequent aid and adaptation measures can further diverge based on the varied circumstances of local residents.

Theories of environmental governance emphasize decision-making and processes, while the environmental state theory I advocate highlights the consequences of these decisions and processes, revealing how environmental policies impact people. This distinction is significant. When policies are explicitly designed to influence society, it is crucial to closely examine their effects on local communities. Welfare, education, and health-related policies serve as prime examples. In contrast, the causal relationships between environmental policies and societal changes are often less clear-cut or visible, making them less likely to be subjects of public debate. The effectiveness of environmental policies is usually assessed based on their impact on environmental quality rather than their effects on people.

I have emphasized that pinpointing when a country evolves into an environmental state is both challenging and essential. Nonetheless, it is vital to recognize that environmental states do not emerge out of a vacuum; they are built upon existing resource governance structures.

Table 7.1
How developmental states become environmental states

	←Developmental state→		←Environmental state→
Interventions / Resources	Production (use is encouraged)	Conservation (use is moderated)	Protection (use is forbidden)
Farmland	Private lands/ land reclamation	Zoning	Enclosure by state
Forests	Deforestation	Sustainable yield through cycles of harvesting and replanting	Establishment of protected areas
Water	Irrigation, transportation, power generation	Usage limits, water management, and improvement of water quality	Protection of water sources
Minerals	Mining	Licensing of mining rights	Shift to renewable resources

Source: Created by the author.

Table 7.1 highlights the key transition in governmental interventions as developmental states adopt the characteristics of environmental states. The state's focus gradually shifts from distant resources to those closely intertwined with ordinary people's daily lives. Policy initiatives evolve from production (where use is encouraged) to conservation (where use is moderated) and protection (where use is prohibited). Throughout this process, policies become increasingly specialized and exclusionary, concentrating power in the hands of experts and the state. Over time, we observe inherent tendencies that push policies in one direction or another, often without those involved being fully aware of the consequences for the affected populations.

There is significant overlap between developmental states and environmental states; what distinguishes them is the degree of penetration of their environmental policies. All countries are currently in a transitional phase between the two. Often, a clear line can be drawn between these stages at the point where nations begin to prioritize nature and sustainability in their planning, before establishing the systems needed for implementation. However, stated intentions and plans rarely reflect the actual extent of state influence. The only definitive way to identify an environmental state is by examining the changes affecting people on the ground. Such impacts can vary greatly over time and in different contexts, complicating efforts to define and identify environmental states objectively.

When we investigate the problem-solving methods of a developmental state, conventional theories of environmental governance become less useful. If we confine our analysis to conservation-related government agencies and other overt state measures, we will struggle to reach the core of the problem. For example, in the context of climate change, states often prefer adaptation over mitigation. Countries pursuing adaptation tend to focus on developing new technologies and constructing infrastructure—from breakwaters to advanced water management systems. While these measures are logical from a development perspective, as they also create jobs, they do little to address the root cause of global warming: rising CO_2 levels (Keohane 2015).

Time and again, states determine that something is lacking or ineffective in a particular area, bring in nonlocal expertise to address the problem, and trigger a reversal of their natural resource and environmental policies. States often implement environmental policies without realizing they are operating within the mindset of a developmental state. In developing countries, the shift from development to conservation can occur very rapidly. Any analysis of these states must consider the systems of governance that existed before the transition from a developmental to an environmental state.

7.4 Choosing Where to Depend

Ideas about competition-based economic self-reliance and the regulatory approaches typically taken by developmental states are ill-suited for environmental management and sustainability. Instead of solely relying on a division of labor that emphasizes exchangeability, we can develop a nuanced network of dependencies that fosters a balance of power. I arrived at this conclusion after exploring the underlying causes of numerous increasingly acute global problems, from terrorism and refugee crises to global warming and inequality. Are these issues not, at least in part, a result of our lingering attachment to the values nurtured by developmental states—namely, competition and self-reliance? Excessive competition obscures the essential conditions for life: cooperation and reliance on others. When systemic disparities and inequalities are rendered invisible, they are easily overlooked and dismissed.

The premise of competition is that resources are scarce, prompting people to fight for what they need from others. Economic anthropologist Karl Polanyi argued long ago that human economies are characterized by their dependence on nature; the assumption of scarcity is a logical twist that only began to make sense *after* countries adopted market economies (Polanyi 1944). Unbeknownst to us, mounting competition has made humans more aggressively dependent on their environment, ultimately disrupting the systems that maintain a balanced relationship between people and nature. The histories of developmental states demonstrate a pattern of dependencies rooted in inequality, subordination, and centralized control. A well-documented example is the policy-driven trend among Latin American countries to specialize in a limited range of commodities, which led to excessive dependence on foreign trade and gave rise to dependency theories in the 1970s (Katz 2023).

Today, it is common to view dependency as a negative condition in which one's autonomy and self-determination are curtailed by subordination to the interests of more powerful entities. The obvious solution for such dependence has historically been independence and self-reliance. However, it soon became clear that, in the context of economic development, independence could also lead to isolation. In response to the nationwide social and health issues related to social isolation, the UK government appointed a Minister of Loneliness in 2018. Despite this awareness of the side effects of self-reliance-focused development, competition and independence continue to dominate our economic landscape.

The history of development is not characterized by grand technological innovations but rather by a gradual rediscovery of how humans depend on one another. A growing focus on gender has illuminated the sex-based division of labor and how gender inequality has become ingrained in society. Similarly, the "discovery" of environmental problems has raised awareness that, contrary to previous beliefs, natural resources are neither free nor endlessly abundant. Increasingly frequent natural disasters reconnect people not only with their dependence on nature but also with the interdependencies between states and communities. Still, the current pinnacle of thinking about development such as the Sustainable Development Goals (SDGs) formulated by the UN and enforced by many governments around the world, scarcely engages with the question of how people should relate to one another. Each society must discover for itself what *good* dependency might look like in its local context. Mutual dependency between the state and the community exists alongside interdependencies among individuals, groups, and society at large.

Is *good* interdependence something we can define? This book has focused on stakeholders who must confront the oppressions imposed by environmental states. The cases I described involve multiple axes of interdependence among a variety of actors, which would be challenging for states that are reliant on extreme control or despotism to navigate. The issue is not the number of stakeholders but rather the relationships among them. Individuals typically belong to multiple societies or systems simultaneously, and communication technology exponentially expands this capacity. The balance among these entities is disrupted only when one becomes so dominant that the system or culture it represents begins to interfere with the other entities to which individuals belong. When the resulting frictions and discrepancies negatively affect communities, inversion follows.

As a prominent scholar of development goals, Amartya Sen proposed that a society can only be just when individuals have the freedom to select the most important identities from the options available to them (Sen 2010). However, many people lack a voice in the institutional frameworks established by their countries, which regulate access to natural resources and the management of the environment. This underscores the importance of fostering a multilayered structure comprising various intermediary organizations—ranging from community unions to citizen networks—that span overlapping regions.

Developmental states built on private property and social contracts prioritize competition and self-reliance in their efforts to overcome dependencies. I argue that environmental states should instead focus on improving the quality of the interdependencies that connect stakeholders. I concur with the view that the quickest way to reduce poverty is to prioritize the *good* interdependencies that ensure people's survival, rather than isolating individuals as mere "productive laborers" (Ferguson 2015). For instance, in Japan, a declining birth rate combined with low immigration will likely continue to increase the aging population, who must depend not only on government aid but also on a limited number of family members. While interdependencies must exist to support individual self-reliance, the success of this support system hinges on how we design a network of mutual reliance.

Historically, many rural areas have maintained a strained relationship with the state, largely characterized by governments extracting resources through forced labor, taxes, or "gifts" of local products.

With the advent of democracy, however, governments began to offer returns in the form of education, health care, and other public services. In other words, the state's dependence on its citizens is evident in the taxes, products, and labor it demands. These domestic interdependencies also underlie competition between nations. The rise of individualism and the invisible spread of the division of labor—based on what individuals can produce and exchange within society—has cultivated a fragmented form of mutual dependence.

Herbert Spencer (1820–1903), best known as the proponent of civilizational progress in the nineteenth century, emphasized the importance of mutual dependence. He argued that societies function similarly to biological organisms, where increasing specialization leads to greater interdependence among individuals and institutions (Spencer 1876). However, as the market economy and liberal individualism gained prominence, his insights were largely forgotten. Why? Perhaps because mutual dependence primarily manifests through the unconscious division of labor, which makes it easy to overlook. After years of prioritizing development over environmental policies, we must reexamine the *types* of dependencies these ideologies have fostered in our societies. It seems an opportune moment to consider whether we should move beyond a dependency framework rooted in the conventional division of labor.

The *unit* of analysis in dependency also warrants scrutiny. Scholars of international politics have devoted significant attention to the interaction of self-reliance and mutual dependence within the context of international affairs (Keohane and Nye 2000). However, as previously noted, these studies typically use the nation as the unit of analysis and often neglect both international movements and domestic interdependencies. In examining the history of environmental states, we find that the efforts of nations to achieve economic independence are closely intertwined with internal mechanisms of dependency that connect the state and society. All aspects of policies aimed at combating climate change—from monitoring to developing technology for climate adaptation—are fundamentally reliant on the relationships between experts and the state.

Latecomer environmental states must balance the demands of environmental protection and economic development. For these states, any nature-focused interventions will have immediate repercussions for the economy. This inversion will persist until we develop a comprehensive understanding of our dependencies on both nature and external economic forces. Societies need to learn to navigate both realms effectively.

7.5 Possible Counterarguments

In the previous chapters, I discussed a wide range of examples of state interference that curtail people's ability to live their normal daily lives. This may lead some readers to conclude that environmental states are simply "bad" and should be avoided. However, it is those affected by the actions of environmental states who will ultimately determine whether these government intrusions have unwelcome effects. This is why the policy recommendations I provide, based on environmental state theory, are focused on retaining—and perhaps enhancing—citizens' abilities to respond to government policies as they see fit.

To further contextualize my argument, it may be useful to highlight some potential criticisms of the inversion thesis. First, my argument may frustrate environmentalists who have been disheartened by the slow progress of environmental policies. Many would argue that protecting nature is impossible without proactive government initiatives. They contend that the harms that state policies have inflicted on some individuals must be considered unavoidable collateral damage for the greater good of environmental preservation. I argue that this perspective reveals a troubling power structure in which the suffering of certain groups is easily accepted—or, completely overlooked.

I agree that, despite the potential risks posed by environmental states, inaction is not an option. Governments cannot passively observe as forests are depleted, citizens suffer from polluted air, landscapes are cluttered with trash, water quality declines, and temperatures continue to rise. There is an element of inversion in standing by while problems escalate when solutions are available. However, this should not overshadow the potential harm of environmental policies, particularly in peripheral regions inhabited by communities that have, for the most part, not benefited significantly from development.

A related counterargument against this book's analysis is that inversion is a necessary evil. Commentators might argue that environmental destruction has reached such dramatic levels that the time for debate has passed. With the survival of humanity at stake, is it not justifiable to implement conservation measures first and address issues of democracy later? This argument is linked to the concept of "environmental authoritarianism," which suggests that authoritarian governments, like those in China and Vietnam, may tackle environmental problems more efficiently (Beeson 2010). While it is true that, in some cases, authoritarian approaches may be more effective than democratic ones, it is essential to conduct a longitudinal analysis of inversion. Such analysis reveals that when states impose rules for the "national interest," they often decide who benefits from these new policies, frequently neglecting the fair treatment of local populations. This oversight tends to exacerbate existing disparities among citizens.

Additionally, some may argue that if conservation is entrusted to local communities, there is a risk that people may pursue misguided goals or employ ineffective methods to achieve them. In the 1990s, for instance, many countries in the region established "community" forests, fisheries, and irrigation projects, but many of them failed to empower communities or to conserve essential resources (Agrawal and Gibson 1999; Kurauchi et al. 2006). In response, I contend that all conservation measures—regardless of their scale—directly impact certain communities, even if these effects are not immediately apparent. Therefore, those who are most affected by changes in environmental and conservation policies deserve a voice in the decisions that shape their futures.

There are several reasons why one of my observations, recounted in the introduction of this book—where climate education was provided to Laotian people who could not read or write—left me feeling uneasy. Local communities received instruction about certain aspects of the relationship between people and nature and were led to believe that they could expect gifts from urban dwellers. However, they did not receive crucial information that could lead to urgently needed improvements in their health care and basic education. I felt that these communities lacked the power to refuse or negotiate outside intervention in their environment, and that this situation was exacerbated by the fact that they were ostensibly being offered "aid." When communities have enough knowledge to make informed choices about their circumstances, and the consequences of those choices are their own, their decisions should be respected—regardless of whether outside experts believe the communities' chosen goals or methods are misguided. The issue lies in environmental states, where most government interventions occur without people in peripheral regions receiving adequate information about what is happening, let alone being presented with meaningful choices.

When the people whose lives are most intertwined with the natural environment are excluded, their trust in the government erodes, and their lack of support renders environmental policies ineffective. An alternative state might cultivate a civil society capable of appropriately scrutinizing future environmental crises. Instead of embracing or dismissing necessary evils, such an approach seeks to avoid creating acute evils in the first place. It is possible, and we should make the effort.

7.6 Dependencies and Beyond

Environmental states emerge when people delegate, whether voluntarily or involuntarily, their duty and right to protect nature to the government. Observing this trend worldwide offers a unique opportunity to reconsider our relationship with the natural environment. The rapid pace of climate and environmental change compels us to act. However, while environmental states claim to act in service of a future in which we all respect our environment, they simultaneously limit the options available to individual citizens. Throughout the previous chapters, we learned that this process alters not only how humans depend on nature but also how we depend on each other.

It is appropriate to conclude this book with my thoughts on the nature of human dependencies. The concept of dependency presupposes a dichotomized worldview that distinguishes between those who depend and those who are depended upon, and who are therefore independent. While the involved agents may shift and change, this dichotomized framework persists. The renewal of dependency necessitates a transformation of this worldview.

The first non-European to win the Nobel Prize in Literature, Indian poet Rabindranath Tagore (1861–1941), was invited to Japan in 1916. His visit coincided with Japan's effort to break free from Western powers and pursue independence. It is worth recalling Tagore's advice to Japan, given in vain as the country embarked on a regrettable trajectory of war and self-destruction in the thirty years following his visit:

> What is still more dangerous for Japan is not this imitation of the outer features of the West but the acceptance of the motivating

> force of Western civilization as her own. . . . I can see her motto, taken from science, "Survival of the fittest," writ "Help yourself, and never heed what it costs to others;" the motto of the blind man, who only believes in what he can touch, because he cannot see. . . . The moral law, which is the greatest discovery of man, is the discovery of this wonderful truth, that man becomes all the truer, the more he realizes himself in others. (Tagore 1916, 78)

Tagore challenges the prevalent dichotomy of "us" and "them," which fuels nationalism and the essentialization of culture (Said 1981). This provides an interesting parallel with the rise of environmental states. The notion of "realizing oneself in others" transcends the conventional human-environment dichotomy and encourages us to question how individuals or governments—that is, various political categories *within* humanity—must be renewed to fulfill everyone's will. After all, debates about climate and the environment are fundamentally discussions about how we depend on one another.

Notes

Introduction

1. The Asian Development Bank and the World Bank together, for example, have invested in Cambodia to update its key infrastructure in agriculture toward climate-proof systems in preparation for droughts and floods (National Council of Sustainable Development 2024).

2. Developmentalism tends to create a comprehensive set of government ministries and agencies, including the one in charge of environmental matters, that enact basic legislation to consolidate laws that affect nature, ratify relevant international conventions, and train dedicated experts (Duit et al. 2016). In these ways, environmental policy is a part of development in the broadest sense.

3. This tendency was substantially rectified by the emergence of political ecology as a social scientific approach to natural resources and the environment (Blaikie 1985).

4. The *Oxford English Dictionary* has the following definition among others under "environment": "The natural world or physical surroundings in general, either as a whole or within a particular geographical area, esp. as affected by human activity."

5. This point is analogous to Rostow's theory of the stages of economic growth where the time needed to move from one stage to another differs between countries with various starting points, so that the policies that promote economic growth in one stage are different from those that promote growth in other stages (Rostow 1960).

Chapter 1

1. As of 2024, there are over seven hundred projects covering more than fifty-seven countries around the world (see https://unfccc.int/topics/land-use/workstreams/redd/what-is-redd).

2. Policymakers face difficulties in understanding and addressing the social and political impacts of carbon pricing. Although the idea of carbon pricing has been around for a long time, it is still challenging to identify where carbon pricing can be effective and where targeted policies may be necessary (Stoll and Mehling 2021).

3. Today, state interests have expanded beyond the confines of our planet. National governments and international forums are now deliberating on the resources and responsibilities associated with cleaning up debris in outer space, which may lead to the emergence of a new type of environmental state devoid of any "local residents."

4. Similar processes are taking place in the field of renewable energy. Geographer Cheryl McEwan reports from South Africa about the spatial implications of renewable energy, where state-designated zones are becoming new territories that allow political and economic actors to exercise authority and commercial power (McEwan 2016).

5. But see also Naidoo et al. (2019), who claim that a negative relation between protected areas and human well-being is unfounded.

6. Similar logic should apply to the state response to COVID-19, such as when to declare a state of emergency or impose lockdown policies.

Chapter 2

1. Of the 1,422 Minamata disease patients in Kumamoto Prefecture, 378 had died by the end of 1980. The first death occurred in 1954, with a peak incidence in 1956 when Minamata disease was officially reported for the first time. The number of deaths increased rapidly after 1972, reaching a second peak in 1976 (Tamashiro et al. 1984).

2. The Ashio copper mine was the site of Japan's first major pollution disaster in the 1880s and the scene of the 1907 miners' riots. This pollution disaster led to the rise of the Japanese environmental movement and the implementation of the 1897 Third Mine Pollution Prevention Order. Additionally, it prompted operational changes in the mine that contributed to the 1907 riots, part of a series of mining disputes that year (Ui 1992).

3. When conflicts of this nature arise between businesses and communities, the state's interests are often clear, particularly in contexts dominated by growth and

profit motives. The primary motivation of the state is to prevent high-profile disputes from emerging.

4. For more details on Ui's life and thoughts, see Ui 2008, compiled by his students and others.

5. Michael Polanyi initially termed this "tacit knowing" (Polanyi 1966). He asserted that *formal* knowledge, which can be articulated, constitutes only a small portion of our total knowledge; conversely, "we can know more than we can tell" (Polanyi 1966, 4).

6. For a critical examination of the issues surrounding the concept of a "third party" discussed in this chapter, see particularly Funabashi (2007).

7. Yuko Fujigaki, a renowned science and technology expert, points out that "in the social theory of science and technology, the knowledge of local residents is always discussed in comparison with the knowledge of experts in science and technology, whereas in environmental sociology, the contrast is rarely brought to the fore, and the knowledge of local residents is overwhelmingly regarded as reliable and legitimate" (Fujigaki 2004, 26). This observation should serve as a wake-up call against the uncritical glorification of local knowledge.

8. Flyvbjerg (2001), articulating Aristotle's insights, sparked controversy among positivist researchers by arguing that the social sciences should actively address the issue of value to improve context-specific judgments, rather than aiming for universal laws like those in the natural sciences. For discussions of these debates, see Schram and Caterino (2006).

9. Tacit knowing and phronesis are not necessarily the same. The former emphasizes observation regarding the input of knowledge, while the latter focuses on the outcomes of that knowledge. However, both concepts share a common understanding concerning the tacit dimensions that are difficult to articulate.

10. Relevant here is the Austrian economist Friedrich Hayek, who pointed out that "the scientific knowledge known to experts is only a small part of the sum of all knowledge." More than half a century ago, Hayek argued that the advantage of a liberal capitalist economy over a socialist planned economy is that the former is better able to capture "knowledge of a particular circumstance of time and place" through information-intensive "prices" (Hayek 1945, 521). However, for society to tap into such knowledge and utilize it as a whole, Hayek noted that the "man on the spot" must be given appropriate discretionary authority, rather than having all plans made centrally. It is on this basis that Hayek argued for the superiority of market mechanisms over a planned economy.

11. In the classic work *Decision Analysis*, Raiffa framed the issue as the "error of the fourth kind"—that is, "solving the right problem too late" (Raiffa 1968, 5).

12. In some cases, people have not silently accepted top-down standardization. For example, as late as the 1920s, the "window and door tax" (a system of taxation based on the number of windows and doors in a building) remained in effect in France to facilitate tax collection. However, it inadvertently prompted a strategy of resistance by encouraging individuals to reduce the number of windows and doors in their buildings (Scott 1998, 47–48).

13. After witnessing the Sanriku tsunami of 1933 (which happened in almost the same place as the one triggered by the Great East Japan Earthquake in 2011), physicist Torahiko Terada questioned why human preparations for tsunamis have always come too late. He identified forgetfulness as the primary reason, describing it as "a natural phenomenon in the human world" (Terada 1933). The Sanriku tsunami, which left three thousand dead or missing, occurred just thirty-seven years after the Sanriku tsunami of June 15, 1896, which had claimed over twenty thousand lives. Even this relatively short period was sufficient for people to forget the pain of past disasters, leading them to return to coastal areas for the sake of convenience.

14. Regarding the potential of storytelling as a method to convey the unspeakable message, see Hirai (2015).

Chapter 3

1. My use of the term "state" follows Ernest Gellner's definition as an "institution or set of institutions specifically concerned with the enforcement of order" (Gellner 1983, 4), with an emphasis on power structures. By "countries," I refer to the general combination of state and society, emphasizing territorial space. While I view the state as an integral part of society, analyzing it separately can provide greater clarity. In late-nineteenth-century Asia, which is my main focus here, states helped shape and were shaped by social groups that had not yet been integrated into the power structure of the nation-state.

2. There are, of course, significant differences between Japan and Thailand that present challenges for any comparative analysis. One such difference is the geopolitical context: Thailand shares national borders with countries that were colonized by Western powers, while Japan is surrounded by sea, where population pressure on natural resources was much higher. Additionally, by the late nineteenth century, levels of centralization in each state varied significantly. Almost all regions of Japan had well-developed road and communication systems and were known and governed, to some extent, by the Shogunate. In contrast, Thailand's king had limited knowledge of the frontier towns and peripheral populations (Bunnag 1977).

3. In keeping with translation traditions, I use "Ministry of the Interior" for *Krasuwan Mahatai* in Thailand and "Ministry of Home Affairs" for Naimushō in Japan. Both ministries were politically significant and shared the similar function of maintaining order within the state.

4. Local facilities include basic infrastructure such as schools, roads, and bridges, which are built with contributions from the local community to help alleviate the financial burden on the central government in promoting modernization.

5. Thongchai (1994, 65), quoting Burney, described the issue of double taxation, where locals refused to pay taxes to both the British and the Chinese tin miners.

Chapter 4

1. When demand for common resources is low, there is little competition among users. However, as the number of users increases, common resources tend to degrade and can even become depleted, as individuals often prioritize short-term private benefits over the long-term public good.

2. "Enclosure" generally refers to land and forests, but it has also targeted other resources, including coastal areas, lakes, pastures, and special economic zones. The enclosure of the Tonle Sap Lake fishing grounds, initially by the French, represented a significant government incursion that had profound consequences for Cambodians.

3. Fisheries were not confined to designated fishing lots and could operate on public fishing grounds. Large and medium-sized fisheries, classified as commercial entities, were prohibited from operating between June and the end of September. However, this restriction did not apply to small fishing operations, such as those supporting individual families, which could operate year-round, provided they did not use certain types of fishing gear.

4. In 2020, there were 225 community fisheries with 94,122 members across 36 villages operating in the 538,739 hectares of Tonle Sap fishing areas (Camille and Lieng 2020).

5. In 2023, the premier remarked during the inauguration ceremony of the Bakheng Water Treatment Plant in Phnom Penh that "hopefully, in the future, Cambodia may see more investment in the oil sector due to oil resources believed to be available beneath the Tonle Sap, mainland, and in the seabed, which we have yet to study and explore" (Vanyuth 2023).

6. Furthermore, Hun Sen asserted that the Tonle Sap fisheries contributed 1.5 million USD to the government's revenue every year. Given that the economy had been maintaining a growth rate of over 6 percent at the time, this amount is relatively small, making Tonle Sap an unlikely target for economic exploitation.

7. According to a June 2020 survey of some five hundred households around the lake: 80 percent of respondents said they felt there was less conflict, 75 percent reported better access to fish, and 60 percent said they believed it had resulted in less illegal activity on the lake (Esterman 2022).

Chapter 5

1. See the report by Suyanto et al. (2009) for a detailed analysis.

2. Pérez-Blanco et al. (2020) reviewed over 230 theoretical and empirical studies and found that in 83.2 percent of the case studies, water conservation technologies intended to increase physical irrigation efficiency result in increased water consumption, thereby decreasing water availability for other uses.

3. It has been demonstrated that the distance of human dwellings from irrigation schemes significantly influences malaria risk among household members (Mangani et al. 2022).

Chapter 6

1. Such decentralization and devolution of authority to communities are often accompanied by a variety of explicit rules. Local rules that previously functioned implicitly are gradually dismantled and incorporated into the state's logic, where formal knowledge prevails.

2. Durkheim pointed to Montesquieu as a pioneer who first highlighted the importance of intermediary groups. In his masterpiece *The Spirit of the Laws*, Montesquieu explored the idea that the monarchy represented the most highly organized social form of his time. He stated that "intermediary, subordinate and dependent powers," such as the nobility, clergy, and parliaments, are essential checks on the monarch's power, preventing despotism by ensuring that authority is distributed across various levels of society (Montesquieu 1977, 20).

3. In premodern Europe, families and churches served as typical intermediary groups. However, with the emergence of large-scale capitalists in the nineteenth century, masters of small-scale handicrafts began to actively form trade associations to protect their craftsmen. Some researchers consider this movement the origin of labor unions (Unwin 1980, 9).

4. It is within this framework that local governments in certain countries may qualify as intermediary groups, while in others, they are considered mere "outposts" of the central government.

5. Ostrom addresses this aspect in her famous "eight design principles" when she states that "rule-making rights of community members are respected by outside authorities" and that "nested tiers, from the lowest level up to the entire interconnected system," are necessary (Ostrom 1990, 108). However, the relationship between local communities and external authorities—arguably the most defining factor in contemporary globalized society—receives less attention in her work.

6. In a class work on social change, Karl Polanyi left the following insightful remark on the virtue of slowing down: "The rate of change is often of no less importance

than the direction of the change itself; but while the latter frequently does not depend upon our volition, it is the rate at which we allow change to take place that well may depend upon us" (Polanyi 1957, 36–37).

7. Even those who have long kept the state at arm's length can find themselves thrust into the spotlight during a calamity. Large-scale natural disasters exert a potent force, compelling the state to disburse compensation or reallocate land to affected citizens. Such arrangements can prompt the state to (re)assess which citizens should qualify for assistance, as painfully observed during decades of petitions and litigation stemming from the occurrence of Minamata disease in Japan (see chapter 2).

8. Durkheim identified the intricate relationship between the state and intermediaries as follows: "Far from being in opposition to the social group endowed with sovereign powers and called more specifically the State, the State presupposes their [secondary groups'] existence: it exists only where they exist. No secondary groups, no political authority—at least, no authority to which this term can apply without being inappropriate" (Durkheim 1957, 83).

9. Even when these individuals form a mass, they remain vulnerable due to their weak horizontal ties; they must still confront the state as individuals (Kornhauser 1960).

10. Documents from the US Secretary of Defense clearly state that development projects aim to "improve production, income, and the standard of living of the villagers targeted for insurgent recruitment." Such improvements will make the population more resistant to insurgent (communist) propaganda and recruitment appeals (Secretary of Defense 1968, 1).

11. REC has long served as a nexus for environmental experts in Central and Eastern Europe, contributing to camaraderie among those addressing environmental issues in the region. This international intermediary group was established with the unique mission of "orienting" state policies in the field of environmental protection rather than advocating specific rules.

12. Financial difficulties led the REC, originally intended to function as an NGO, to draw closer to the government, stripping it of its autonomous role as an intermediary group and ultimately resulting in the organization's dissolution in 2019.

13. It is important to note that the abundance of land available to migrants from Europe to the United States helped secure the material means for each local unit to become economically self-reliant. The first embassy from Japan, the Iwakura Mission, also found the same in the United States in 1872: "The country possesses the prime requirements for agriculture: good climate and a vast amount of land. . . . Finding their own country too restrictive and its politics too repressive, they came to America to exploit the land and to indulge their yearning for freedom. These people educated the less cultured immigrants, gave them a sense of community, and directed their energies toward opening up the land" (Kume 2009 [1872], 70–71).

References

Abernethy, Charles L. 2011. "Historical Developments in Irrigation Governance." *Proceedings of the Institutions of Civil Engineers* 164 (2): 87–98.

Adas, Michael. 2001. *Agricultural and Pastoral Societies in Ancient and Classical History*. Temple University Press.

Agrawal, Arun. 2005. *Environmentality: Technologies of Government and the Making of Subjects*. Duke University Press.

Agrawal, Arun, and Clark C. Gibson. 1999. "Enchantment and Disenchantment: The Role of Community in Natural Resource Conservation." *World Development* 27 (4): 629–649.

Aklin, Michaël, and Johannes Urpelainen. 2014. "The Global Spread of Environmental Ministries: Domestic—International Interactions." *International Studies Quarterly* 58 (4): 764–780.

Alomoto, W., A. Niñerola, and L. Pié. 2022. "Social Impact Assessment: A Systematic Review of Literature." *Social Indicators Research* 161: 225–250. https://doi.org/10.1007/s11205-021-02809-1.

Ang, Wei Jing, Edward Park, Dung Duc Tran, Ho Huu Loc, and Yasu Pokhrel. 2024. "Dams in the Mekong: A Comprehensive Database, Spatiotemporal Distribution, and Hydropower Potentials." *Earth System Science Data* 16 (3): 1209–1228. https://doi.org/10.5.494/essd-16-1209-2024.

Ang, Wei Jing, Edward Park, Yadu Pokhrel, Dung Duc Tran, and Ho Huu Loc. 2024. "Dams in the Mekong: A Comprehensive Database, Spatiotemporal Distribution, and Hydropower Potentials." *Earth System Science Data Discussions* 16: 1209–1228. https://doi.org/10.5194/essd-16-1209-2024.

Aristotle. 1954. *The Nicomachean Ethics of Aristotle*. Translated and introduced by Sir David Ross. Oxford University Press.

Aristotle. 2004. *The Metaphysics*. Penguin Classics.

Ascher, William. 1999. *Why Governments Waste Natural Resources*. Johns Hopkins University Press.

ASEAN Secretariat. 2023. Sixth ASEAN State of the Environment Report. Accessed December 5, 2023. https://asean.org/wp-content/uploads/2023/09/SOER6_Layout_0920_ISSN_sent.pdf.

Asian Development Bank. 2019. *Lao People's Democratic Republic Energy Sector Assessment, Strategy, and Road Map*. Asian Development Bank.

Attwood, Donald W. 1987. "Irrigation and Imperialism: The Causes and Consequences of a Shift from Subsistence to Cash Cropping." *Journal of Development Studies* 23 (3): 341–366.

Attwood, Donald W., and B. S. Baviskar. 1987. "Why Do Some Co-Operatives Work but Not Others? A Comparative Analysis of Sugar Co-Operatives in India." *Economic and Political Weekly* 22 (26): A38–A56. http://www.jstor.org/stable/4377145.

Baeza, Andres, Menno J. Bouma, Andy P. Dobson, Ramesh Dhiman, Harish C. Srivastava, and Mercedes Pascual. 2011. "Climate Forcing and Desert Malaria: The Effect of Irrigation." *Malaria Journal* 10 (July): 190.

Bahadur, Krishna, Vittoria Elliott, Ratha Seng, Robert S. Pomeroy, Jared Schenkels, and Evan D. G. Fraser. 2020. "Evaluating Community Fishery Management Using Fishers' Perceptions in the Tonle Sap Lake of Cambodia." *Environmental Development* 33 (100503).

Baran, Eric, and Un Borin. 2012. "The Importance of the Fish Resource in the Mekong River and Examples of Best Practices." In *From Sea to Source; International Guidance for the Restoration of Fish Migration Highways*, edited by Peter Philipsen, Peter Paul Schollema, and Herman Wanningen, 136–141. Regional Water Authority Hunze en Aa's Postbus.

Barker, Randolph, and F. François Molle. 2004. *Evolution of Irrigation in South and Southeast Asia*. Research Report 5. Comprehensive Assessment Secretariat.

Barton, Gregory A., and Brett M. Bennett. 2010. "Forestry as Foreign Policy: Angle-Siamese Relations and the Origins of Britain's Informal Empire in the Teak Forests of Northern Siam, 1883–1925." *Itinerario* 34 (2): 65–86.

Beck, Ulrich. 1998. "Politics of Risk Society. In *The Politics of Risk Society*, edited by Jane Franklin, 9–22. Polity.

Beeson, Mark. 2010. "The Coming of Environmental Authoritarianism." *Environmental Politics* 19 (2): 276–294.

Berlin, Sir Isaiah. 1953. *The Hedgehog and the Fox*. Weidenfeld & Nicolson.

Berlin, Sir Isaiah. 1996. *The Sense of Reality*. Random House UK.

Bissonnette, Jean-François, and Rodolphe De Koninck. 2017. "The Return of the Plantation? Historical and Contemporary Trends in the Relation Between Plantations and Smallholdings in Southeast Asia." *Journal of Peasant Studies* 44 (4): 918–938.

Black, Simon, Antung A. Liu, Ian W. H. Parry, and Nate Vernon-Lin. 2023. *IMF Fossil Fuel Subsidies Data: 2023 Update*. IMF.

Blaikie, Piers. 1985. *The Political Economy of Soil Erosion in Developing Countries*. Longman.

Bodin, Örjan. 2017. "Collaborative Environmental Governance: Achieving Collective Action in Social-Ecological Systems." *Science* 357 (6352): 659.

Boehm, Hans-Dieter, and Hasjrul Junaid. 2015. The Mega Rice Project, Central Kalimantan, Indonesia SKEPHI-1998. https://doi.org/10.13140/RG.2.1.2883.9845.

Booth, Anne. 2013. "Colonial Revenue Policies and the Impact of the Transition to Independence in South East Asia." *Bijdragen tot de Taal-, Land- en Volkenkunde* 169 (1): 37–67.

Bresser-Pereira, Luiz Carlos. 2019. "Models of the Developmental State." *CEPAL Review* 128: 35–48

Brockington, Daniel, and James Igoe. 2006. "Eviction for Conservation: A Global Overview." *Conservation and Society* 4 (3): 424–470.

Brummelhuis, Han ten. 2005. *Kings of the Waters: Homan van der Heide and the Origin of Modern Irrigation in Siam*. Silkworm Books.

Bruna, Natacha. 2022. "A Climate-Smart World and the Rise of Green Extractivism." *Journal of Peasant Studies* 49 (84): 839–864.

Bruna, Natacha. 2023. *The Rise of Green Extractivism: Extractivism, Rural Livelihoods and Accumulation in a Climate-Smart World*. Routledge.

Bunnag, Tej. 1977. *Provincial Administration of Siam, 1892–1915: The Ministry of the Interior Under Prince Damrong Rajanubhab*. Oxford University Press.

Busby, Joshua. 2018. "Warming World: Why Climate Change Matters More than Anything Else." *Foreign Affairs* 97 (4): 49–57.

Camille, Bann, and Sopha Lieng. 2020. "Fish Counts: Increasing the Visibility of Small-Scale Fisheries (SSFs) in Cambodia's National Planning." *IIED Working Paper*. International Institute for Environment and Development.

Castree, Noel. 2008a. "Neoliberalising Nature: The Logics of Deregulation and Reregulation." *Environment and Planning A: Economy and Space* 40 (1): 131–152.

Castree, Noel. 2008b. "Neoliberalising Nature: Processes, Effects, and Evaluations." *Environment and Planning A: Economy and Space* 40 (1): 153–173.

Chang, Ha-Joon. 2002. *Kicking Away the Ladder: Development Strategy in Historical Perspective*. Anthem Press.

Charoenmuang, Tanet. 2006. *Thailand: A Late Decentralizing Country*. Urban Development Institute Foundation.

Cooke, Nola. 2011. "Tonle Sap Processed Fish." In *Chinese Circulations/Capital, Communities, and Networks in Southeast Asia*, edited by Eric Tagliacozzo and Wen-chin Chang, 360–379. Duke University Press.

Cooperman, Michael S., Nam So, Mauricio E. Arias, Thomas A. Cochrane, Vittoria Elliott, Taber Hand, Lee Hannah, et al. 2012. "A Watershed Moment for the Mekong: Newly Announced Community Use and Conservation Areas for the Tonle Sap Lake May Boost Sustainability of the World's Largest Inland Fishery." *Cambodian Journal of Natural History* 2 (December): 101–106.

Coward, E. Walter. 1980. "Irrigation Development: Institutional and Organizational Issues." In *Irrigation and Agricultural Development in Asia: Perspectives from the Social Sciences*, edited by E. Walter Coward, 15–27. Cornell University Press.

Coyle, Diane. 2015. *GDP: A Brief but Affectionate History*, rev. and expanded ed. Princeton University Press.

Cullather, Nick. 2000. "Development? It's History." *Diplomatic History* 24 (4): 641–653.

Cushman, Jennifar W. 1991. *Family and State: The Formation of a Sino-That Tin-Mining Dynasty; 1797–1932*. Oxford University Press.

Degen, Peter, Frank Van Acker, Niek van Zalinge, Nao Thuok, and Ly Vuthy. 2000. "Taken for Grant: Conflict over Cambodia's Freshwater Fish Resources." Paper presented at the International Association for the Study of Common Property.

Dina, Thol, and Jin Sato. 2014. "Is Greater Fishery Access Better for the Poor? Explaining De-Territorialisation of the Tonle Sap, Cambodia." *Journal of Development Studies* 50 (7): 962–976.

Ding, David K., and Sarah E. Beh. 2022. "Climate Change and Sustainability in ASEAN Countries." *Sustainability* 14 (2): 999. https://doi.org/10.3390/su14020999.

Doner, F. Richard, Bryan K. Ritchie, and Dan Slater. 2005. "Systemic Vulnerability and the Origins of Developmental States: Northeast and Southeast Asia in Comparative Perspective." *International Organization* 59: 327-361.

Dorn, Felix Malte, Robert Hafner, and Christina Plank. 2022. "Towards a Climate Change Consensus: How Mining and Agriculture Legitimize Green Extractivism in Argentina." *Extractive Industries and Society* 11 (September): 101–130.

Dove, Michael. 1993. "A Revisionist View of Tropical Deforestation and Development." *Environmental Conservation* 20 (1): 17–24.

Dove, Michael. 2012. "A Political-Ecological Heritage of Resource Contest and Conflict." In *Routledge Handbook of Heritage in Asia*, edited by Patrick T. Daly and Tim Winter, 182–198. Routledge.

Dreyfus, Hubert, and Paul Rabinow. 1982. *Michel Foucault: Beyond Structuralism and Hermeneutics*. 2nd ed. University of Chicago Press.

Duit, Andreas, Peter H. Feindt, and James Meadowcroft. 2016. "Greening Leviathan: The Rise of the Environmental State?" *Environmental Politics* 25 (1): 1–23.

Dunbar, Robin. 2022. *How Religion Evolved and Why It Endures*. Penguin Books.

Durkheim, Émile. 1957. *Professional Ethics and Civic Morals*. Routledge & Kegan Paul.

Durkheim, Émile. 1974. *Shakaigaku Kōgi* [Lectures on Sociology]. Translated by Miyajima Takashi and Kawakita Takashi. Misuzushobō.

Embree, John. 1950. "Thailand: A Loosely Structured Social System." *American Anthropologist* 52 (2): 181–193.

Enomoto, Kazuhiro, and Ishikawa Satoshi. 2008. "Tonresā puko no Suisan shigen to kan'ri–Suisan sigen kan'ri no mokuteki to kadai ni tsuite" [Fishery Resources and their Management in Tonle Sap]. In *Hito to Sakana no sizen'shi–haha naru mekon'gawa ni ikiru* [Natural History of Fish and People], edited by Akimichi Tomoya and Kurokura Hisashi, 201–219. Sekaisisōsha.

Eriksen, Siri, E. Lisa F. Schipper, Morgan Scoville-Simonds, Katharine Vincent, Hans Nicolai Adam, Nick Brooks, Brian Harding, et al. 2021. "Adaptation Interventions and Their Effect on Vulnerability in Developing Countries: Help, Hindrance or Irrelevance?" *World Development* 141 (May): 105–383.

Esterman, Isabel. 2022. "Fisheries Crackdown Pushes Cambodians to the Brink on Tonle Sap Lake." *Mongabay News* (blog). Newstex. August 26.

Fairhead, James, Melissa Leach, and Ian Scoones. 2012. "Green Grabbing: A New Appropriation of Nature?" *Journal of Peasant Studies* 39 (2): 237–261.

Falcus, Malcom. 1989. "Early British Business in Thailand." In *British Business in Asia Since 1860*, edited by R. P. T. Davenport-Hines and Geoffery Jones, 117–156. Cambridge University Press.

FAO (Food and Agriculture Organization of the United Nations). 1993. "The State of Food and Agriculture." FAO Agriculture Series no. 26.

FAO. 2023. "Land Use Statistics and Indicators 2000–2021." Global, Regional and Country Trends. FAOSTAT Analytical Briefs Series no. 71.

FAO. 2011. *The State of the World's Land and Water Resources for Food and Agriculture (SOLAW): Managing Systems at Risk.* Food and Agriculture Organization of the United Nations, Rome and Earthscan, London.

Feeny, David. 1989. "Decline of Property Rights in Man in Thailand, 1800–1913." *Journal of Economic History* 49 (2): 285–296.

Ferguson, James. 2015. *Give a Man a Fish: Reflections on the New Politics of Distribution.* Duke University Press.

Flynn, Gerald, and Vutha Srey. 2022. "Fisheries Crackdown Pushes Cambodians to the Brink on Tonle Sap Lake." MONGABAY. August 26. https://pulitzercenter.org/stories/fisheries-crackdown-pushes-cambodians-brink-tonle-sap-lake

Flyvbjerg, Bent. 2001. *Making Social Science Matter: Why Social Inquiry Fails and How It Can Succeed Again.* Cambridge University Press.

Foucault, Michel. 2009. *Security, Territory, Population: Lectures at the College De France, 1977–78.* Springer.

Franco, Jennifer C., and Saturnino M. Borras. 2019. "Grey Areas in Green Grabbing: Subtle and Indirect Interconnections Between Climate Change Politics and Land Grabs and Their Implications for Research." *Land Use Policy* 84: 192–199.

Freeman, David M., and Max K. Lowdermilk. 1991. "Middle-Level Farmer Organizations as Links Between Farms and Central Irrigation Systems." In *Putting People First: Sociological Variables in Rural Development,* edited by Michael M. Cernea, 108–144. World Bank Group.

Freire, Paulo. 1970. *Pedagogy of the Oppressed.* Seabury Press.

Fujigaki, Yūko. 2004. "Kagaku gijutsu shakai ron (STS) to kankyō shakai gaku no setten" [Connection Between Science and Technology Studies and Environmental Sociology]. *Kankyō shakai gaku kenkyū* 10: 25–41.

Fukushima, Masao. 1970. *Chisokaisei no kenkū* [Land Tax Reform of Meiji Era]. Yūhikaku.

Funabashi Harutoshi. 2007. "Ui Jun no shigoto no shakaigaku he no shisa to kyori" [Ui Jun's Work and Their Implication and Distance to Sociology]. *Kankyō shakaigaku kenkyū* 13: 233–238.

Furnivall, John Sydenham. 1956. *Colonial Policy and Practice: A Comparative Study of Burma and Netherlands India.* New York University Press.

Garon, Sheldon. 1987. *The State and Labor in Modern Japan.* University of California Press.

Geertz, Clifford. 1963. *Agricultural Involution: The Process of Ecological Change in Indonesia.* University of California Press.

Global Carbon Project. 2024. "Global Carbon Project 2023." Accessed April 4, 2024. https://www.globalcarbonproject.org/index.htm.

Global Witness. 2009. *Country for Sale: How Cambodia's Elite Has Captured the Country's Extractive Industries*. Global Witness.

Gopnik, Alison. 2023. "Caregiving in Philosophy, Biology & Political Economy." *Daedalus* 152 (1): 58–69.

Gordon, Scott H. 1954. "The Economic Theory of a Common-Property Resource: The Fishery." *Journal of Political Economy* 62 (2): 124–142.

Guha, Ramachandra. 2000. *Environmentalism: A Global History*. Longman.

Haber, Stephen, and Victor Menaldo. 2011. "Do Natural Resources Fuel Authoritarianism? A Reappraisal of the Resource Curse." *American Political Science Review* 105 (1): 1–26.

Hall, Derek, Philip Hirsch, and Tania Murray Li. 2011. *Powers of Exclusion: Land Dilemmas in Southeast Asia*. NUS Press.

Hardin, Garrett. 1968. "The Tragedy of Commons." *Science* 162 (3859): 1243–1248.

Hayasaka, Keizō. 2013. "Kōkōgyō no iwate shinshutsu to gyogyō iriai to no shōtotsu -suiri ken wo meguru tanaka kōzan seitetsujo tai dogawa sakeryū gyogyō no soshou jirei" [Progression of Mining Industries to Iwate Prefecture and their Conflicts with Fishery Commons]. *Artes Liberales* 91: 13–35.

Hayek, Friedrich. A. 1945. "The Use of Knowledge in Society." *American Economic Review* 35 (4): 519–530.

Hendry, Joy. 2014. *Science and Sustainability Learning from Indigenous Wisdom*. Palgrave Macmillan New York.

Hirai, Kyōnosuke. 2015. "Storytelling as Political Practice: Habitus and Social Change in the Minamata Disease Movement." *Senri Ethnological Studies* 91 (July): 81–99.

Hirayama, Naoko. 2016. "Suiro karyū no mizubusoku to mizu kan'ri ni kakawaru komyunikēshon no jittai" [Water Shortages in the Downstream and the Reality of Communication Regarding Water Administration]. In *Mizu wo wakatsu–Chīki no mirai kanousei no kyōsō*, edited by Kubota Jyunpei, 87–103. Benseishuppan.

Hirschman, Albert. 1958. *Strategies of Economic Development*. Yale University Press.

Hirschman, Albert. 1967. *Development Projects Observed*. Brookings.

IEA. 2024. "Southeast Asia Energy Outlook 2024." Accessed March 20, 2024. https://www.iea.org/reports/southeast- asia-energy-outlook-2024.

Ingram, James C. 1971. *Economic Change in Thailand, 1850–1970*. Stanford University Press.

Inoue, Tatsuo. 2001. *Gendai no hinkon* [Today's Problem of Poverty]. Iwanamishoten.

Ishi, Yoneo. 1978. "History and Rice-Growing." In *Thailand: A Rice Growing Society*, edited by Yoneo Ishii, 15–39. Monographs of the Center for Southeast Asian Studies, Kyoto University, no. 12. University of Hawaii Press.

Ishikawa, Satoshi, Mina Hori, Akira Takagi, Thuok Nao, Kazuhiro Enomoto, and Hisashi Kurokura. 2008. "Historical Changes on the Fisheries Management in Cambodia." *Tropics* 17 (4): 315–324.

Jamin, Mohammad, and Sapto Hermawan. 2024. "Valuing Indigenous Rights over Forest Governance in Six ASEAN Member Countries." *UUM Journal of Legal Studies* 15 (2): 585–621.

Japan International Cooperation Agency. 2016. Biribiri kangai jigyou hy*ōka hōkokusho* [Evaluation Report of the Bili-Bili Irrigation Project]. Japan International Cooperation Agency.

Japan Times. 2024. "Japan Court Rejects Redress Claims by Unrecognized Minamata Victims," *Japan Times*, March 22.

Johnson, Chalmers. 1982. *MITI and the Japanese Miracle*. Stanford University Press.

Jones, Julia P. G., and Simon L. Lewis. 2023. "Forest Carbon Offsets Are Failing." *Science* 381 (6660): 830–831.

Jones, Sam. 2015. "Net Loss: Fish Stocks Dwindle in Cambodia's Tonlé Sap Lake." *Guardian*, December 1. https://www.theguardian.com/global-development/2015/dec/01/cambodia-tonle-sap-lake-fish-stocks-dwindle-conservation.

Kasai, Toshiyuki. 2003. "Kanbojia Tonresappuko chīki no kankyō hozen ni tsuite no yobi teki kōsatsu" [A Preliminary Observation on the State of Environmental Conservation in Tonle Sap Lake Areas]. *Ristumeikan Kokusai chīki kenkyū* 21: 41–64.

Kato, Tsuyoshi. 2014. "'Kaihatsu' gainen no seisei wo megutte—shogen kara shokuminchi shugi no jidai made" [On the Evolution of Development Concept]. *Asian and African Area Studies* 13 (2): 112–147.

Katz, Claudio. 2023. *Dependency Theory After Fifty Years: The Continuing Relevance of Latin American Critical Thought*. Haymarket Books.

Keohane, Robert O. 2015. "Global Politics for Climate Change: Challenge for Political Science." *Political Science & Politics* 48 (1): 19–26.

Keohane, Robert O., and Joseph Nye Jr. 2000. *Power and Interdependence*. Cambridge University Press.

Keohane, Robert O., and Elinor Ostrom. 1995. *Local Commons and Global Interdependence: Heterogeneity and Cooperation in Two Domains*. Sage.

Kikuchi, Michiki. 1981. "Hogo ryō shihai kakuritsu ki no Kambojia no naisuimen gogyō" [Inland Fisheries During the Protectorate Period of Cambodia]. *Hitotsubashi ronsō* 86 (4): 497–524.

Kikuma, Mitsuru. 1980. "Kokuyū rin'ya no jimoto riyō to ikurin rōdō soshiki no tenkaikōzō –itaku rin seido no shiteki bunseki" [Local Use of State Forests and the Evolution of Labor Organization for Silviculture]. *Hokkaidō daigaku nōgakubu en'shūrin ken'kyū hōkoku* 37 (2): 479–608.

Kinzley, Judd C. 2018. *Natural Resources and the New Frontier: Constructing Modern China's Borderlands*. University of Chicago Press.

Kitahara, Atsushi. 2010. "Shinrin no hozen to riyō no mujun to funsō" [Conflicts and Contradictions in the Conservation and Utilization of Forests]. In *Funsō kaiketsu gurōbaruka chīki bunka*, edited by Pauline Kent and Kitahara Atsushi, 63–84. Mineruvashobō.

Kitahara, Atsushi. 2012. *Tai kindai tochi shin'rin seisaku shi kenkyū* [Historical Study on Land and Forest Policies in Thailand]. Kōyōshobō.

Knaus, Christopher. 2020. "One Million Cambodians Under Threat from Development of Vital Wetlands: Report," *The Guardian*, July 27, 2020. https://www.theguardian.com/world/2020/jul/27/one-million-cambodians-under-threat-from-development-of-vital-wetlands-report

Kobayashi, Mitsue. 1968. *Kokuyūchi iriaiken no kenkyū* [Study of Communal Rights in State Lands]. University of Tokyo Press.

Kobayashi, Naoki. 2021. "Minamata no neko wo meguru kioku" [Memories of a Cat in Minamata]. *Hosei Journal of Sociology and Social Sciences* 67 (49): 69–95.

Kobori Satoru. 2017. "Rinkai kaihatsu, kōgai taisaku, shizen hogo–kōdoseichōki Yokohama no kankyō shi" [Coastal Development, Pollution Abatement, and Resource Conservation]. In *Sengo nihon no Kaihatsu to min'shusuhgi—chīki ni miru sōkoku*, edited by Shōji Shunsaku, 71–104. Shōwadō.

Kornhauser, William. 1960. *The Politics of Mass Society*. Routledge and Kegan Paul.

Krom Sapayakon Taranii. 1992 *100 pi krom sapayakon tarani* [100 years of the Department of Mines]. Chalongratrana Publishing.

Kume, Kunitake. 2009 [1872]. *Japan Rising: The Iwakura Embassy to the US and Europe 1871–1873*. Cambridge University Press.

Kurauchi, Yuko, Antonio La Vina, Nathan Badenoch, and Lindsey Fransen. 2006. *Decentralization of Natural Resources Management: Lessons from Southeast Asia,*

Synthesis of Decentralization Case Studies Under the Resources Policy Support Initiative (REPSI). World Resources Institute.

Kuroda, Yōichi, and François Nectoux. 1989. *Nettairin hakai to nihon no mokuzai bōeki* [Destruction of Tropical Rainforests and the Japanese Timber Trade]. Tsukijishokan.

Kurumisawa, Yoshiki. 2004. "Chīki chūkan'dan'tai ni yoru shizen shigen no ijikan'ri." In *Kankyō to hō*, edited by Ushiyama Tsumoru, Sutō Shigeyuki, Ōtsuka Tadashi, Suami Takao, and Kurumisawa Yoshiki, 139–176. Seibundō.

Kusumastuti, Dyah Indriana, Vera Chania Putri, Dwi Jokowinarno, and Endro Prasetyo Wahono. 2021. "Farmers' Participation in Irrigation Management in the Punggur Utara Irrigation Area, Indonesia." *Pertanika Journal: Science & Technology* 29 (2): 725–741.

Lamont, C. Hempel. 1996. *Environmental Governance: Global Challenge*. Island Press.

Lansing, J. Stephen. 1991. *Priests and Programmers: Technologies of Power in the Engineered Landscape of Bali*. Princeton University Press.

Lansing, J. Stephen. 2007. *Priests and Programmers: Technologies of Power in the Engineered Landscape of Bali*. Princeton University Press.

Loos, Tamara. 1994. "Introduction." In *Five Years in Siam from 1891–1896*, edited by H. Warington Smyth, 5–19. White Lotus.

Lorenzen, Stephan, and Rachel P. Lorenzen. 2008. "Institutionalizing the Informal: Irrigation and Government Intervention in Bali." *Development* 51 (1): 77–82.

Mabry, Bevars D. 1979. *The Development of Labor Institutions in Thailand*. Data Paper no. 112, Southeast Asia Program. Cornell University.

Macaulay, R. H. 1934. *History of the Bombay Burmah Trading Corporation, Ltd. 1864–1910*. Spottiswoode, Ballantyne.

Maemura, Yu. 2024. "*Jijodoryoku*: The Spirit of Self-Help in Development Cooperation." *The Semantics of Development in Asia*, edited by Jin Sato and Kim Soyeun. Springer Nature: 129–148.

MAFF (Ministry of Agriculture, Forestry and Fisheries). 2015. *Agricultural Sector Strategic Development Plan 2014–2018*. MAFF.

Magtolis, Carlos M., Jr., and Jojema D. Indab. 2008. "A Policy Study on Environmental Protection in the Philippines." *Philippine Journal of Public Administration* 52 (2–4): 364–379.

Mahood, Simon P., Colin M. Poole, James E. M. Watson, Richard A. MacKenzie, Sahadev Sharma, and Stephen T. Garnett. 2020. "Agricultural Intensification Is

Causing Rapid Habitat Change in the Tonle Sap Floodplain, Cambodia." *Wetlands Ecology and Management* 28 (5): 713–726.

Mahr, Dana. 2021. *The Knowledge of Experience*, Palgrave Macmillan.

Malloch, D. E. 1852. *Siam: Some General Remarks on Its Productions and Particularly on Its Imports and Exports, and the Mode of Transacting Business with People.* J. Thomas at the Baptist Mission Press.

Mann, Michael. 1984. "The Autonomous Power of the State: Its Origins, Mechanisms and Results." *European Journal of Sociology* 25 (2): 185–213. https://doi.org/10.1017/S0003975600004239.

Mansfield, Cristina, and Kurt MacLeod. 2004. *Commune Councils and Civil Society: Promoting Decentralization Through Partnerships*. Pact Cambodia.

Martinez-Alier, Joan. 2014. "The Environmentalism of the Poor." *Geoforum* 54 (July): 239–241.

Maruyama, Masao. 1968. "Patterns of Individuation and the Case of Japan: A Conceptual Scheme." In *Changing Japanese Attitudes Toward Modernization*, edited by Marius B. Jansen, 489–531. Princeton University Press.

Mase, Tōru. 1984. *Mizu ga tsukutta ajia* [Asia Created by Water]. Ienohikarikyōkai.

Matsubara, Nozomu. 2002. "Kankyō gaku ni okeru dēta no jūbunsei to ishiketteihandan" [Completeness of Data and Decision Making in Environmental Studies]. In *Kankyōgaku no gihō*, edited by Ishi Hiroyuki, 167–214. University of Tokyo Press.

Matsunami, Hidemi. 1920. *Meiji rin'gyou shiyō* [Forestry Documents from the Meiji Period]. Dainippon san'rin kai.

Matsushima, Shizuo. 1978. *Tomoko no shakaigaku teki kōsatsu: kōzan rōdōsha no itonamu kyōdōseikatsutai bunseki* [Sociological Investigation to Tomoko]. Ochanomizushobō.

McEwan, Cheryl. 2017. "Spatial Processes and Politics of Renewable Energy Transition: Land, Zones and Frictions in South Africa." *Political Geography* 56 (January): 1–12.

Mekvichai, Banasopit. 1988. "The Teak Industry in Northern Thailand: The Role of Natural Resource-Based Regional Economy." PhD diss., Department of City and Regional Planning, Cornell University.

Milne, Sarah, and Sango Mahanty. 2015. *Conservation and Development in Cambodia: Exploring Frontiers of Change in Nature, State and Society*. Routledge. https://doi.org/10.4324/9781315887302.

Ministry of the Environment. 2020. *Annual Report on Environmental Statistics*. https://www.env.go.jp/doc/toukei/tokeisyu.

Ministry of Trade and Industry. 1966. *Shōkōseisakushi*, Vol. 22 [History of Commerce and Industrial Policy]. Edited by Ministry of Trade and Industry.

Mitchell, Timothy. 2011. *Carbon Democracy: Political Power in the Age of Oil*. Verso.

Mizuno Kōichi. 1981. *Tai nōson no shakai soshiki* [Social Organization of Rural Thailand]. Sōbunsha.

Mol, Arthur P. J. 2016. "The Environmental Nation State in Decline." *Environmental Politics* 25 (1): 48–68.

Molle, François, Peter P. Mollinga, and Philippus Wester. 2009. "Hydraulic Bureaucracies and the Hydraulic Mission: Flows of Water, Flows of Power." *Water Alternatives* 2 (3): 328–349.

Montesquieu. *The Spirit of Laws: A Compendium of the First English Edition*. Edited by David Wallace Carrithers. University of California Press, 1977.

Montgomery, John D. 1983. "When Local Participation Helps." *Journal of Policy Analysis and Management* 3 (1): 90–105. https://doi.org/10.2307/3324007.

Morita, Masaru. 2012. *Chikasui wa kataru: mienai shigen no kiki* [The story of the underground water: The crisis of invisible resources]. Iwanami Shoten.

MRC. 2022. *Joint Environmental Monitoring Programme at Two Mekong Mainstream Dams: The Don Sahong and Xayaburi Hydropower Projects*. MRC Secretariat. https: //doi .org/10.52107/mrc.aqrs7o.

Murai, Yoshinori, and ODA Chōsa kenkyūkai. 1989. *Musekinin ODA taikoku nippon* [Japan as an Irresponsible Major Power of ODA] Takarajimasha.

Murakushi, Nisaburō. 1999. *Nihon no dentōteki rōshi kankei* [Traditional Labor Relations in Japan]. Sekaishoin.

Mutert, Earnest, and Thomas Fairhurst. 2002. Developments in Rice Production in Southeast Asia. *Better Crops International* 15: 12–17.

Mushiake, Katumi. 2002. "Hydrology and Water Resources in Monsoon Asia." *Journal of Japanese Society of Hydrology and Water Resources* 15 (4): 428–434.

Naidoo, Robin, Drew Gerkey, Dave Hole, Alexander Pfaff, Alicia M. Ellis, Christopher D. Golden, Diego Herrera, et al. 2019. "Evaluating the Impacts of Protected Areas on Human Well-Being Across the Developing World." *Science Advances* 5 (4): eaav3006.

Nakashima, Masahiro. 2003. "Mizushigen Kaihatsu to hozen no aida [Between Water Resources Development and Water Conservation]" In *Mizu no bunka 13*, edited by Mizkan mizu no bunka sentā, 10–13. Mizkan mizu no bunka sentā.

National Archives in Cambodia. National Archive File No. 24105, Phnom Penh.

National Council of Sustainable Development. 2024. "Projects on Climate Change." https://ncsd.moe.gov.kh/dcc/project/enhancement-flood-and-drought-management-pursat-province-under-greater-mekong-sub-region-flood-and-drought-risk-management-and-mitigation.

Nem Singh, J. 2022. "The Renaissance of the Developmental State in the Age of Post Neoliberalism." In *Handbook of Governance and Development*, edited by Wil Hout and Jane Hutchison, 97–114. Edward Elgar.

Nevins, Joseph, and Nancy Lee Peluso. 2008. *Taking Southeast Asia to Market: Commodities, Nature, and People in the Neoliberal Age*. Cornell University Press.

Nishio, Takashi. 2021. *Nihon shinrin gyōseishi no kenkyū: Kankyō hozen no genryū* [A Study of Japanese Forestry Administration History]. University of Tokyo Press.

Niwa, Kunio. 1989. *Tochimondai no kigen* [The Origins of Land Issues]. Heibonsha.

Nixon, Rob. 2011. *Slow Violence and the Environmentalism of the Poor*. Harvard University Press.

Nkoa, Bruno Emmanuel Ongo, Jacques Simon Song, and Brice Minkoue Bikoua. 2024. "Natural Resource Rents in Developing Countries: Is the Positive Influence on the Fragilities Real?" *Resources Policy* 89 (February): 104541. https://doi.org/10.1016/j.resourpol.2023.104541.

Noddings, Nei. 1984. *Caring: A Feminine Approach to Ethics and Moral Education*. University of California Press.

O'Brien, Thomas. 2010. "Stuck in the Middle: Maintaining the Organizational Legitimacy of the Regional Environmental Center." *Voluntas: International Journal of Voluntary and Non-Profit Organizations* 21 (March): 339–357.

OECD. 2023a. "Policy Instruments for the Environment (PINE) Database." http://oe.cd/pine.

OECD. 2023b. *Climate Finance Provided and Mobilised by Developed Countries in 2013–2021: Aggregate Trends and Opportunities for Scaling Up Adaptation and Mobilised Private Finance*. Climate Finance and the USD 100 Billion Goal, OECD Publishing. https://doi.org/10.1787/e20d2bc7-en.

Ogino, Toshio. 1990. *Nihon Kindairinsei no Hattatsu Katei* [Development Processes of Japanese Modern Forest Policy]. Ringyō chōsakai.

Oguni, Kazuko. 2016. "Mizukanri wo meguru hitobito no kachi no yukue—minami suraueshi shū biribiri kan'gai jueki chiku ni okeru kyōdōteki jissen ni mukete" [Directions of Value Systems Among Local people with Regard of Water Management]. In *Mizu wo wakatsu–Chīki no mirai kanousei no kyōsō*, edited by Kubota Jyunpei, 35–58. Benseishuppan.

Öjendal, Joakim, and Mona Lilja. 2009. *Beyond Democracy in Cambodia: Political Reconstruction in a Post-Conflict Society* NIAS Press.

Öjendal, Joakim, and Kim Sedara. 2011. *Real Democratization in Cambodia? An Empirical Review of the Potential of a Decentralization Reform.* Swedish International Center for Local Democracy (ICLD).

Omar, Padam Jee, and Vikram Kumar. 2022. "Assessment of Damage for Dam Break Incident in Lao PDR Using SAR Data." *International Journal of Hydrology Science and Technology* 14 (4): 421–434.

Ongo, Nkoa, Bruno Emmanuel, Jacques Simon Song, and Brice Minkoue Bikoula. 2024. "Natural Resource Rents in Developing Countries: Is the Positive Influence on the Fragilities Real?" *Resources Policy* 89 (February): 104541.

Ostrom, Elinor. 1990. *Governing the Commons: The Evolution of Institutions for Collective Action.* Cambridge University Press.

Ostrom, Elinor. 1994. "Constituting Social Capital and Collective Action." In *Local Commons and Global Interdependence: Heterogeneity and Cooperation in Two Domains,* edited by Robert O. Keohane and Elinor Ostrom, 126–160. SAGE Publications.

Ōta, Takehiko. 2012. *Shinrin hōwa—kokudo no henbō wo kangaeru* [Saturating Forests]. NHK Books.

Partelow, Stefan, Achim Schlüter, Derek Armitage, Maarten Bavinck, Keith Carlisle, Rebecca Gruby, Anna-Katharina Hornidge, et al. 2020. "Environmental Governance Theories: A Review and Application to Coastal Systems." *Ecology and Society* 25 (4). https://doi.org/10.5751/ES-12067-250419.

Peluso, Nancy, and Peter Vandergeest. 2001. "Genealogies of the Political Forest and Customary Rights in Indonesia, Malaysia, and Thailand." *Journal of Asian Studies* 60 (3): 761–812.

Peluso, Nancy, and Peter Vandergeest. 2020. "Writing Political Forests." *Antipode* 52 (4): 1083–1103.

Peou, Sorpong. 2007. "Limits of Economic Assistance." *International Democracy Assistance for Peacebuilding,* 180–192. Palgrave Macmillan.

Pérez-Blanco, C. Dionisio, Arthur Hrast-Essenfelder, and Chris Perry. 2020. "Irrigation Technology and Water Conservation: A Review of the Theory and Evidence." *Review of Environmental Economics and Policy* 14 (2): 216–239.

Polanyi, Karl. 1944. *The Great Transformation.* Rinehart.

Polanyi, Michael. 1966. *The Tacit Dimension.* University of Chicago Press.

Porntney, Ken. 2015. *Sustainability.* MIT Press.

Prime Minister Outlines Cooperation Measures to Protect Fisheries Resources. 2012. *Catch and Culture* 18 (2): 30–31.

Pritchett, Lant, and Michael Woolcock. 2004. "Solutions When the Solution Is the Problem: Arraying the Disarray in Development." *World Development* 32 (2): 191–212.

Puangthong, Pawakapan. 2021. *Infiltrating Society: The Thai Military's Internal Security Affairs*. ISEAS Publishing.

Pudjiastuti, T. N., and S. C. M. Wong. 2023. "The Politics of Forced Migration in Southeast Asia." In *Migration in Southeast Asia*, edited by S. Petcharamesree and M. P. Capaldi. IMISCOE Research Series. Springer. https://doi.org/10.1007/978-3-031-25748-3_12.

Pye Oliver. 2023. "Civil Society and Environmentalism: Crossing Frontiers of Activism" In *Routledge Handbook of Civil and Uncivil Society in Southeast Asia*, edited by M. L. Weiss and Hansson E., 329–345. Routledge.

Qadir M, Wichelns D, Raschid-Sally L, McCornick PG, Drechsel P, Bahri Aet al. 2010. "The challenges of wastewater irrigation in developing countries." *Agricultural Water Management* 97: 561–568.

Radkau, Joachim. 2008. *Nature and Power: A Global History of the Environment*. Translated by Thomas Dunlap. Cambridge University Press.

Raiffa, Howard. 1968. *Decision Analysis: Introductory Lectures on Choices Under Uncertainty*. Addison-Wesley.

Ramsay, Ansil. 1979. "Modernization and Reactionary Rebellions in Northern Siam." *Journal of Asian Studies* 38 (2): 283–297.

Ratner, Blake D. 2006. "Policy Review Community Management by Degree? Lessons from Cambodia Fisheries Reform." *Society and Natural Resources* 19 (1): 79–86.

Ratner, Blake D., Guy Halpern, and Mam Kosal. 2011. "Catalyzing Collective Action to Address Natural Resource Conflict: Lessons from Cambodia's Tonle Sap Lake." CAPRi Working Paper 103. https://doi.org/10.2499/CAPRiWP103.

Ratner, Blake D., Sovannarith So, Kosal Mam, Il Oeur, and Sour Kim. 2017. "Conflict and Collective Action in Tonle Sap Fisheries: Adapting Governance to Support Livelihoods." *Natural Resources Forum* 41 (April): 71–82.

Rea, Christopher M., and Scott Frickel. 2023. "The Environmental State: Nature and the Politics of Environmental Protection." *Sociological Theory* 41 (3): 255–281.

Redfern, S. K., N. J. Azzu, and S. Binamira. 2012. "Rice in Southeast Asia: Facing Risks and Vulnerabilities to Respond to Climate Change. Building Resilience for

Adaptation to Climate Change in the Agriculture Sector," *FAO/OECD Workshop Proceedings*, April 23–24, 2012, 295–314.

Repetto, Robert, and Malcolm Gillis. 1988. *Public Policies and the Misuse of Forest Resources*. Cambridge University Press.

Resource Council. 1960. "Suishitsu odaku bōshi taisaku ni taisuru chōsa hōkoku" [Research Report on the Policies to Address Water Pollution]" *Kagaku gijutsu chō shigen chōsakai hōkoku* 15.

Ricks, Jacob I. 2015. "Pockets of Participation: Bureaucratic Incentives and Participatory Irrigation Management in Thailand." *Water Alternatives* 8 (2): 193–214.

Ricks, Jacob I. 2017. "Street-Level Bureaucrats and Irrigation Policy Reform in Southeast Asia." *Asian Politics and Policy* 9 (2): 310–319.

Ross, Corey. 2014. "The Tin Frontier: Mining, Empire, and Environment in Southeast Asia, 1870s–1930s," *Environmental History* 19 (July): 454–479.

Rostow, Walt Whitman. 1960. *The Stages of Economic Growth: A Non-communist Manifesto*. Cambridge University Press.

Sachs, Jeffrey, and Pia Malaney. 2002. "The Economic and Social Burden of Malaria." *Nature* 410 (February): 680–685.

Sagardoy, Juan A., Anthony Bottrall, and G. O. Uittenbogaard. 1982. "Organization, Operation and Maintenance of Irrigation Schemes." FAO Irrigation and Drainage Paper 40. Food and Agriculture Organization.

Said, Edward W. 1981. *Covering Islam*. Vintage Books.

Saitō, Osamu. 2014. *Kankyō no keizaishi -shin'rin, shijō, kokka*. Iwanamishoten.

Sampath, Rajan K. 1992. "Issues in Irrigation Pricing in Developing Countries." *World Development* 20 (7): 967–977.

Sasaki, Tōru. 1971. "Waga kuni no shoki kōgyō rōdō hogo rippou ni tsuite—kōgyō jōrei no kōhu hogo kitei ni kan'suru oboegaki" [Early Legal Protection Measures for Mining Laborers in Japan]. In *Shakaikagakunenpō*, edited by the Institute for Social Science Senshu University, 5. Miraisha.

Sasaoka, Masatoshi. 2019. "Kankyō gabanansu no 'shinten' niyoru Minzokuchi no muryokuka" [The Invalidation of Tacit Local Knowledge Caused by the "Improvement" of Institutional Arrangements for Environmental Governance]. *Bulletin of the Graduate School of Letters, Hokkaido University* 74–89.

Sato, Jin. 2000. "People in Between: Conservation and Conversion of Forest Lands in Thailand." *Development and Change* 31 (1): 155–177.

Sato, Jin. 2003. "Public Land for the People: The Institutional Basis of Community Forestry in Thailand." *Journal of Southeast Asian Studies* 34 (2): 329–346.

Sato, Jin. 2013a. "State Inaction in Resource Governance: Natural Resource Control and Bureaucratic Oversight in Thailand." In *Governance of Natural Resources*, edited by Sato Jin, 15–41. United Nations University Press.

Sato, Jin. 2013b. "Kindaika to tōchi no bunka -meiji nihon to shamu no ten'nenshigen kanri" [Modernization and the Culture of Governance]. In *Kokusai bunka kankeishi kenkyū*, edited by Hirano Ken'ichirō, Furuta Kazuko, Tsuchida Akio, and Kawamura Yōko, 171–192. University of Tokyo Press.

Sato, Jin. 2019. "Fukai tōchi" [Deppe Governance]. In *Shigen kankyō seisaku no keisei katei*, edited by Terao Tadayoshi, 157–171. Institute of Developing Economies.

Sato Jin. 2021. *Kaihatsukyōryoku no tsukurarekata–jiritsu to izon no seitaishi* [The Making of Development Cooperation in Japan]. University of Tokyo Press.

Saunders, F. P. 2014. "The Promise of Common Pool Resource Theory and the Reality of Commons Projects." *International Journal of the Commons* 8: 636–656.

Scheidel, Arnim, and Courtney Work. 2018. "Forest Plantations and Climate Change Discourses: New Powers of 'Green' Grabbing in Cambodia." *Land Use Policy* 77 (September): 9–18.

Schoenberger, Laura. 2017. "Struggling Against Excuses: Winning Back Land in Cambodia." *Journal of Peasant Studies* 44 (4): 870–890.

Schoenberger, Laura, Derek Hall, and Peter Vandergeest. 2017. "What Happened When the Land Grab Came to Southeast Asia?" *Journal of Peasant Studies* 44 (4): 697–725.

Schönweger, Oliver, Andreas Heinimann, Michael Epprecht, Juliet Lu, and Palikone Thalongsengchanh. 2012. *Concessions and Leases in the Lao PDR: Taking Stock of Land Investments. Centre for Development and Environment (CDE).* University of Bern, Geographica Bernensia.

Schram, Sanford F., and Brian Caterino. 2006. *Making Political Science Matter: Debating Knowledge, Research, and Method.* NYU Press.

Scott, James. 1998. *Seeing Like a State: How Certain Schemes to Improve the Human Conditions Have Failed.* Yale University Press.

Scott, James. 2009. *The Art of Not Being Governed.* Yale University Press.

Secretary of Defense. 1968. "Development Organizations and Programs." In *Counterinsurgency Organizations and Programs in Northeast Thailand*, Vol. 5, *Research*

Analysis Corporation for the Join Thai-US Military Research and Development Center by Office of the Secretary of Defense. Advanced Research Projects Agency.

Seiff, Abby. 2017. "When There Are No More Fish." *Eater*. https://eater.com/2017/12/29/16823664/tonle-sap-drought-cambodia.

Sen, Amartya. 2010. *The Idea of Justice*. Belknap Press.

Sethakul, Ratanaporn. 1989. "Political, Social, and Economic Changes in the Northern States of Thailand Resulting from the Chiang Mai Treaties of 1874 and 1883." PhD diss., Department of History, Northern Illinois University.

Shafik, Minouche. 2021. *What We Owe Each Other: A New Social Contract for a Better Society*. Princeton University Press.

Shang, Baoping. 2021. "The Poverty and Distributional Impacts of Carbon Pricing: Channels and Policy Implications." IMF Working Paper WP/21/172.

Shigetomi, Shin'ichi. 1996. *Tai nōson no Kaihatsu to jūmin soshiki* [Village Organizations for Rural Development in Thailand]. Institute of Developing Economies.

Shigetomi, Shin'ichi. 1997. "Tai nōson no 'kyōyū chi' ni kan'suru tochi seido" In *Tōnanajia no keizaikaihatsu to tochiseido* [Economic Development and Land System in South-East Asia], edited by Mizuno Kōsuke and Shigetomi Shin'ichi, 263–302. Ajia Kēzai Kenkyū Jo.

Shiva, Vandana. 2007. *Biopiracy: The Plunder of Nature and Knowledge*. North Atlantic Books.

Shiva, Vandana. 2015. *Earth Democracy: Justice, Sustainability, and Peace*. North Atlantic Books.

Shivakoti, Binaya Raj, and Pham Ngoc Bao, eds. 2020. *Environmental Changes in Tonle Sap Lake and Its Floodplain: Status and Policy Recommendations*. Institute for Global Environmental Strategies (IGES), Tokyo Institute of Technology (Tokyo Tech) and Institute of Technology of Cambodia (ITC).

Siebert, Stefan, Verena Henrich, Karen Frenken, and Jacob Burke. 2013. *Update of the Digital Global Map of Irrigation Areas to Version 5*. Institute of Crop Science and Resource Conservation, Rheinische Friedrich-Wilhelms-Universität.

Sithirith, Mak. 2011. "Political Geography of Tonle Sap: Power, Space, and Resources." PhD thesis, Faculty of Arts and Social Science, The National University of Singapore.

Sithirith, Mak. 2014. "The Patron-Client System and Its Effect on Resources Management in Cambodia: A Case in the Tonle Sap Lake." *Asian Politics & Policy* 6 (4): 595–609.

Sithirith, Mak. 2016. "Political Economy of Fishing Villages: A Case Study in the Tonle Sap Lake, Cambodia." *Journal of Environmental Science and Engineering B* 5 (6): 299–313.

Sithirith, Mak. 2017. "Water Governance in Cambodia: From Centralized Water Governance to Farmer Water User Community." *Resources* 6 (3): 44.

Sithirith, Mak. 2022. "Analysing Transboundary Water Governance: A Global-Regional-National Water Governance Framework of Tonle Sap Lake." *Lakes & Reservoirs: Science, Policy and Management for Sustainable Use* 27 (1): e12396.

Skinner, William. 1957. *Chinese Society in Thailand: An Analytical History*. Cornell University Press.

Slade, Herbert. 1897. *Report of the Forestry Department*. National Archive of Thailand, Bangkok. File R5 M 16/9 (Ministry of the Interior).

Slade, Herbert. 1901. *Report of the Royal Forestry Department for 119* (M5 16/3). Chotmaihet Rachakan Thi 5, Krasuang Mahathai. National Archives of Thailand.

Slocomb, Margaret. 2004. "Commune Elections in Cambodia: 1981 Foundations and 2002 Reformulations." *Modern Asian Studies* 38 (2): 447–467.

Smith, Adam. 1976 [1776]. *An Inquiry into the Causes of the Wealth of Nations*. University of Chicago Press.

Smith, Eugene W., and Aileen M. Smith. 1975. *Minamata: Words and Photographs by W. Eugene Smith and Aileen M. Smith*. Alskong-Sensoriam.

Smith, Rhona. 2021. "Relocating Floating Communities in Cambodia: Kampong Chhnang." *Human Rights Quarterly* 43 (2): 290–312.

Smyth, H. Warington. 1898. *Five Years in Siam, from 1891–95: The Malay and Cambodian Peninsulars, with Descriptions of Ruby Mines*. White Lotus Press.

Sok, Serey, Fenglong Wang, and Nyda Chhinh. 2021. "Political Participation and Small-Scale Fishery Management in the Tonlé Sap, Cambodia." *International Journal of Water Resources Development* 38 (4): 616–633.

Sokhem, Pech, and Kengo Sunada. 2006. "The Governance of the Tonle Sap Lake, Cambodia: Integration of Local, National and International Level." *Water Resources Development* 22 (3): 399–416.

Spencer, Herbert. 1876–1896. *The Principles of Sociology*. 3 vols. Williams and Norgate.

Stoll, Christian, and Michael A. Mehling. 2021. "Climate Change and Carbon Pricing: Overcoming Three Dimensions of Failure." *Energy Research & Social Science* 77 (July): 102062.

Stott, Philip. 1992. "Angkor: Shifting the Hydraulic Paradigm." In *The Gift of Water: Water Management, Cosmology and the State in South East Asia*, edited by Jonathan Rigg, 47–58. School of Oriental and African Studies, University of London.

Suehiro, Akira. 2000. Kyacchiappugata *kōgyōka ron–ajia keizai no kiseki to tenbō* [Catch-Up Industrialization: The Trajectory and Prospects of East Asian Economies]. University of Nagoya Press.

Sugimoto, Ayumi. 2013. "Chizu wo meguru chi no kōryū—Firipin no sank agata shinrin seisaku wo jirei tosihte" [Intellectual Exchanges Through Maps]. *Kankyō shakaigaku kenkyū* 19 (November): 96–111.

Sugiura, Mikiko. 2008. "Kangai yōsui no kankō ni narau–'kishōka' shita shigen no bunpai mekanizumu" [Learning from Custamary Irrigation Water Use]. In *Shigen wo miru me -genba kara no bunpai ron*, edited by Jin Sato, 148–164. Tōshindō.

Suhardiman, Diana. 2018. "Linking Irrigation Development with the Wider Agrarian Context: Everyday Class Politics in Water Distribution Practices in Rural Java." *Journal of Development Studies* 54 (3): 413–425.

Sumi, Kazuo. 1988. *ODA enjo no genjitsu* [The Realities of ODA]. Iwanamishoten.

Suwa, Masaru. 1996. *Hakai nippon ODA 40 nen no tsumeato* [The Destructive Japan: 40 Years of ODA]. Aokishoten.

Suyanto, Noviana Khususiyah, Idris Sardi, Yana Buana, and Meine van Noordwijk. 2009. "Analysis of Local Livelihoods from Past to Present in the Central Kalimantan Ex-mega Rice Project Area." Working paper 94. World Agroforestry Centre.

Tachibanaki, Toshiaki. 2017. *Kakei no keizaigaku* [Economics of Household]. Iwanamishoten.

Tagore, Rabindranath. 1916. *The Spirit of Japan*. Maven Books.

Takahashi, Akio. 1996. "Kami biruma chausē chihou no kasen kangai to nōgyō" [River Irrigation and Agriculture in the Northern Burma of Kyaukse]. In *Ajia no kangai seido -mizu riyō no kōristuka ni mukete*, edited by Horī Kenzō, Shinoda Takashi, and Tada Hirokazu, 169–214. Shinhyōron.

Takayanagi, Tomohiko. 2017. "Shinrin shigen to tochi shoyu" [Forest Resources and Land Tenure]. In *Keizai shakai no rekishi*, edited by Nakanishi Satoru, 66–85. University of Nagoya Press.

Tamashiro, Hidehiko, Hirokatsu Akagi, Mikio Arakaki, Makoto Futatsuka, and Lewis H. Roht. 1984. "Causes of Death in Minamata Disease: Analysis of Death Certificates." *International Archives of Occupational and Environmental Health* 54 (April): 135–146.

Tambunlertchai, Somsak, and S. P. Gupta. 1993. *Development Planning in Asia*. Asian and Pacific Development Centre.

Tanabe, Shigeharu. 1981. "Peasant Farming Systems in Thailand: A Comparative Study of Rice Cultivation and Agricultural Technology in Chiangmai and Ayutthaya." PhD diss., School of Oriental and African Studies, University of London, ProQuest (10731461).

Tang, Zhen, Hamid Fauziah, Ismail Yusoff, and Vivian Chan. 2023. "A Review of PFAS Research in Asia and Occurrence of PFOA and PFOS in Groundwater, Surface Water and Coastal Water in Asia." *Ground Water for Sustainable Development* 22 (3): 100947.

Taylor, Robert H. 2009. *The State in Myanmar*. Hurst.

Terada, Torahiko. 1933. "Tsunami to Ningen" [Tsunami and People]. Recompiled in Terada Torahiko zenshū dai nana kan [the 7th volume of the *Collected Essays of Terada Torahiko*], 1997. Iwanamishoten.

Terao, Tadayoshi. 2009. "Shigen riyō no rigai chōsei toshite no suishitsu hozen seisaku-minamatabyō jiken to suisitsu nihō wo chūshin ni" [Water Quality Conservation as Method to Reconsolidate Conflicts of Interests]. In *Keizai hatten katei ni okeru kankyō shigen hozen seisaku no keisei*, edited by Terao Tadayoshi, 27–39. Institute of Developing Economies.

Thompson, Virginia. 1947. *Labor Problems in Southeast Asia*. Yale University Press.

Thongchai, Winichakul. 1994. *Siam Mapped: A History of the Geo-Body of a Nation*. University of Hawaii Press.

Thurbon, Elizabeth, Sung-Young Kim, Hao Tan, and John Mathews. 2023. *Developmental Environmentalism: State Ambition and Creative Destruction in East Asia's Green Energy Transition*. Oxford University Press.

de Tocqueville, Alexis. 1863. *Democracy in America*. Vol. 1. Translated by Henry Reeve. Sever and Francis.

Tomosugi, Takashi. 1976. "Tai no kangai nōgyō" [Irrigation Agriculture in Thailand]. In *Ajia no kangai nōgyō*, edited by Fukuda Hitoshi, 117–152. Institute of Developing Economies.

Tönnies, Ferdinand. 1957. *Community and Society* [Gemeinschaft und Gesellschaft]. Michigan State University Press.

Torigoe, Hiroyuki. 1989. "Keiken to seikatsu kankyō shugi" [Experience and the Everyday Life-Environmentalism]. In *Kankyōmondai no shakai riron*, edited by Torigoe Hiroyuki, 14–53. Ochanomizushobō.

Totman, Conrad. 1998. *The Green Archipelago: Forestry in Preindustrial Japan*. University of California Press.

Tsurumi, Kazuko. 1998. "Hyōhaku to Teijū to" [On Drifting and Settling]. In *Tsurumi Kazuko mandara IV*, 240–274. Chikumashobō.

Ui, Jun. 1971. *Kōgai genron* [Principle of Pollution]. Akishobō.

Ui, Jun. 1992. *Industrial Pollution in Japan*. United Nations University Press.

Ui, Jun. 2000. "Kōgai ni okeru chi no kōyō" [Utilities of Knowledge in Pollution Problems]. In *Gensetsu: kirisaku*, edited by Kurihara Akira and Komori Yōichi, 49–72. University of Tokyo Press.

Ui, Jun. 2014. *Genten toshite no minamatabyō* [Minamata as a Starting Point], edited by Fujibayashi Yasushi, Miyauchi Taisuke, and Tomozawa Yūki. Shinsensha.

Ui, Noriko. 2008. *Aru kōgai kankyō gakusha no ashidori: Tsuitō ui jun ni Manabu* [Footsteps of a Pollution Scholar]. Akishobō.

Vandergeest, Peter, and Nancy Lee Peluso. 1995. "Territorialization and State Power in Thailand." *Theory and Society* 24 (3): 385–426.

Van der Meer, Canute. 1988. "Dependent Irrigation Systems in Southeast Asia: Complications for Irrigation Performance Improvements." *Agricultural Administration and Extension* 30 (4): 281–291.

Vanyuth, Chea. 2023. "PM Calls for Study on Cambodia's Untapped Oil Reserves." *Khmer Times*, June 20. https://www.khmertimeskh.com/501310756/pm-calls-for-study-on-cambodias-untapped-oil-reserves/.

Vikrom, Mathur, and Mak Sithirith,. 2008. "Entitlements and the Community Fishery in the Tonle Sap: Is the Fishing Lot System Still an Option for Inland Fisheries Management?" Fisheries Action Coalition Team's paper. https://www.researchgate.net/publication/350063632_Entitlements_and_the_Community_Fishery_in_The_Tonle_Sap_Is_the_Fishing_Lot_System_Still_an_Option_for_Inland_Fisheries_Management

Vong, Rylida. 2019. "Effect of Community-Based Natural Resource Management on Conservation and Poverty Reduction: Evidence from Tonle Sap Lake, Cambodia." PhD diss., Nagoya University, no. 12943.

Wade, Robert. 2018. "The Developmental State: Dead or Alive?" *Development and Change* 49 (2): 518–546

Wang, Xinxin, and Kevin Lo. 2021. "Just transition: A conceptual review," *Energy Research & Social Science* 82.

Wales, H. G. Quaritch. 1934. *Ancient Siamese Government and Administration*. Bernard Quaritch.

Walzer, Michael. 1983. *Spheres of Justice: A Defense of Pluralism and Equality*. Basic Books.

Watanabe, Takashi. 2014. *Hyakushō tachi no mizushigen Sensō–edo jidai no mizu arasoi wo ou*. [Water Resource Wars of the Peasants]. Sōshishbunko.

Wittfogel, Karl. 1957. *Oriental Despotism: A Comparative Study of Total Power.* Yale University Press.

Wolters, Willem. 2007. "Geographical Explanations for the Distribution of Irrigation Institutions: Cases from Southeast Asia." In *A World of Water: Rain, Rivers and Seas in Southeast Asia*, edited by Peter Boomgaard, 209–234. KITLV Press.

Yang, Ruby, and Thomas Lennon. 2011. "The Warriors of Qiugang." Yale Environment 360, Video. https://e360.yale.edu/features/the_warriors_of_qiugang_a_chinese_village_fights_back.

Yoshiki, Fumio. 1980. *How Japan's Metal Mining Industries Modernized.* United Nations University.

Yoshimura, Chihiro, Rajendra Khanal, and Uk Sovannara, eds. 2022. *Water and Life in Tonle Sap Lake.* Springer Nature.

Yoshinaga, Fusako, and Gavan McCormack. 2004. "Minamata: The Irresponsibility of the Japanese State." *Japan Focus* 2 (12).

Yu, Wansik, Yeonsu Kim, Daeeop Lee, and Giha Lee. 2019. "Hydrological Assessment of Basin Development Scenarios: Impacts on the Tonle Sap Lake in Cambodia." *Quaternary International* 503 (part A): 115–127.

Zafra-Calvo, Noelia, Eneko Garmendia, Unai Pascual, Ignacio Palomo, Nicole Gross-Camp, Daniel Brockington, Jose-Antonio Cortes Vazquez, et al. 2019. "Progress Toward Equitably Managed Protected Areas in Aichi Target 11: A Global Survey." *BioScience* 69 (3): 191–197.

Zhang, Ming, Ying Yang, Panpan Du, Jiancheng Wang, Yaoyao Wei, Jiayan Qin, and Limei Yu. 2024. "The Effect of Public Environmental Participation on Pollution Governance in China: The Mediating Role of Local Governments' Environmental Attention." *Environmental Impact Assessment Review* 104 (January): 1073452024

Index

Note: Figures and tables are indicated by italicized page numbers.

American and Comparative Environmental Policy

David M. Konisky, Megan Mullin, Kate O'Neill, series editors

Russell J. Dalton, Paula Garb, Nicholas P. Lovrich, John C. Pierce, and John M. Whiteley, *Critical Masses: Citizens, Nuclear Weapons Production, and Environmental Destruction in the United States and Russia*

Daniel A. Mazmanian and Michael E. Kraft, eds., *Toward Sustainable Communities: Transition and Transformations in Environmental Policy*

Elizabeth R. DeSombre, *Domestic Sources of International Environmental Policy: Industry, Environmentalists, and U.S. Power*

Kate O'Neill, *Waste Trading among Rich Nations: Building a New Theory of Environmental Regulation*

Joachim Blatter and Helen Ingram, eds., *Reflections on Water: New Approaches to Transboundary Conflicts and Cooperation*

Paul F. Steinberg, *Environmental Leadership in Developing Countries: Transnational Relations and Biodiversity Policy in Costa Rica and Bolivia*

Uday Desai, ed., *Environmental Politics and Policy in Industrialized Countries*

Kent Portney, *Taking Sustainable Cities Seriously: Economic Development, the Environment, and Quality of Life in American Cities*

Edward P. Weber, *Bringing Society Back In: Grassroots Ecosystem Management, Accountability, and Sustainable Communities*

Norman J. Vig and Michael G. Faure, eds., *Green Giants?: Environmental Policies of the United States and the European Union*

Robert F. Durant, Daniel J. Fiorino, and Rosemary O'Leary, eds., *Environmental Governance Reconsidered: Challenges, Choices, and Opportunities*

Paul A. Sabatier, Will Focht, Mark Lubell, Zev Trachtenberg, Arnold Vedlitz, and Marty Matlock, eds., *Swimming Upstream: Collaborative Approaches to Watershed Management*

Sally K. Fairfax, Lauren Gwin, Mary Ann King, Leigh S. Raymond, and Laura A. Watt, *Buying Nature: The Limits of Land Acquisition as a Conservation Strategy, 1780–2004*

Steven Cohen, Sheldon Kamieniecki, and Matthew A. Cahn, *Strategic Planning in Environmental Regulation: A Policy Approach That Works*

Michael E. Kraft and Sheldon Kamieniecki, eds., *Business and Environmental Policy: Corporate Interests in the American Political System*

Joseph F. C. DiMento and Pamela Doughman, eds., *Climate Change: What It Means for Us, Our Children, and Our Grandchildren*

Christopher McGrory Klyza and David J. Sousa, *American Environmental Policy, 1990–2006: Beyond Gridlock*

John M. Whiteley, Helen Ingram, and Richard W. Perry, eds., *Water, Place, and Equity*

Judith A. Layzer, *Natural Experiments: Ecosystem-Based Management and the Environment*

Daniel A. Mazmanian and Michael E. Kraft, eds., *Toward Sustainable Communities: Transition and Transformations in Environmental Policy, second edition*

Henrik Selin and Stacy D. VanDeveer, eds., *Changing Climates in North American Politics: Institutions, Policymaking, and Multilevel Governance*

Megan Mullin, *Governing the Tap: Special District Governance and the New Local Politics of Water*

David M. Driesen, ed., *Economic Thought and U.S. Climate Change Policy*

Kathryn Harrison and Lisa McIntosh Sundstrom, eds., *Global Commons, Domestic Decisions: The Comparative Politics of Climate Change*

William Ascher, Toddi Steelman, and Robert Healy, *Knowledge and Environmental Policy: Re-Imagining the Boundaries of Science and Politics*

Michael E. Kraft, Mark Stephan, and Troy D. Abel, *Coming Clean: Information Disclosure and Environmental Performance*

Paul F. Steinberg and Stacy D. VanDeveer, eds., *Comparative Environmental Politics: Theory, Practice, and Prospects*

Judith A. Layzer, *Open for Business: Conservatives' Opposition to Environmental Regulation*

Kent E. Portney, *Taking Sustainable Cities Seriously: Economic Development, the Environment, and Quality of Life in American Cities, second edition*

Raul Lejano, Mrill Ingram, and Helen Ingram, *The Power of Narrative in Environmental Networks*

Christopher McGrory Klyza and David J. Sousa, *American Environmental Policy: Beyond Gridlock, updated and expanded edition*

Andreas Duit, ed., *State and Environment: The Comparative Study of Environmental Governance*

Joseph F. C. DiMento and Pamela Doughman, eds., *Climate Change: What It Means for Us, Our Children, and Our Grandchildren, second edition*

David M. Konisky, ed., *Failed Promises: Evaluating the Federal Government's Response to Environmental Justice*

Leigh Raymond, *Reclaiming the Atmospheric Commons: The Regional Greenhouse Gas Initiative and a New Model of Emissions Trading*

Robert F. Durant, Daniel J. Fiorino, and Rosemary O'Leary, eds., *Environmental Governance Reconsidered: Challenges, Choices, and Opportunities, second edition*

James Meadowcroft and Daniel J. Fiorino, eds., *Conceptual Innovation in Environmental Policy*

Barry G. Rabe, *Can We Price Carbon?*

Kelly Sims Gallagher and Xiaowei Xuan, *Titans of the Climate: Explaining Policy Process in the United States and China*

Matto Mildenberger, *Carbon Captured: How Business and Labor Control Climate Politics*

Mary Alice Haddad, *Effective Advocacy: Lessons from East Asia's Environmentalists*

George Hoberg, *The Resistance Dilemma: Place-Based Movements and the Climate Crisis*

Roger Karapin and David Vogel, *When Federal Climate Policy Works: Benefits, Business, and Politics*

Jin Sato, *Sustainability Inverted: How Environmental Policies Control People*

Joshua Basseches, *Owning the Green Grid: The Political Economy of Renewable Energy Policy Design*

Christopher McGrory Klyza, *Building the American Green State: The States, the Federal Government, and Intercurrence, 1800–2000*

Geoffrey Henderson, *Getting to Work on Climate Policy: The Strategy of Labor-Environmental Coalitions*

Publisher contact:
The MIT Press
Massachusetts Institute of Technology
77 Massachusetts Avenue, Cambridge, MA 02139
mitpress.mit.edu

EU Authorised Representative:
Easy Access System Europe, Mustamäe tee 50,
10621 Tallinn, Estonia
gpsr.requests@easproject.com

Printed by Integrated Books International,
United States of America